ILLUSTRATED GLOSSARY
FOR SOLAR AND SOLAR-TERRESTRIAL PHYSICS

ASTROPHYSICS AND SPACE SCIENCE LIBRARY

A SERIES OF BOOKS ON THE RECENT DEVELOPMENTS
OF SPACE SCIENCE AND OF GENERAL GEOPHYSICS AND ASTROPHYSICS
PUBLISHED IN CONNECTION WITH THE JOURNAL
SPACE SCIENCE REVIEWS

VOLUME 69

ILLUSTRATED GLOSSARY FOR SOLAR AND SOLAR-TERRESTRIAL PHYSICS

edited by

A. BRUZEK and C. J. DURRANT
Fraunhofer Institut, Freiburg

D. REIDEL PUBLISHING COMPANY
DORDRECHT-HOLLAND/BOSTON-U.S.A.

Library of Congress Cataloging in Publication Data

Main entry under title:

Illustrated glossary for solar and solar-terrestrial
physics.

(Astrophysics and space science library; v. 69)
Bibliography: p.
Includes index.
1. Sun. 2. Cosmic physics. I. Bruzek, Anton. II. Durrant,
Christopher J., 1942– III. Series.
QB521.144 523.7 77-11087

ISBN-13:978-94-010-1247-8 e-ISBN-13:978-94-010-1245-4
DOI: 10.1007/978-94-010-1245-4

Published by D. Reidel Publishing Company,
P.O. Box 17, Dordrecht, Holland

Sold and distributed in the U.S.A., Canada, and Mexico
by D. Reidel Publishing Company, Inc.
Lincoln Building, 160 Old Derby Street, Hingham,
Mass. 02043, U.S.A.

Softcover reprint of the hardcover 1st edition 1977

The authors and editors
dedicate this book
to the memory of

K. O. KIEPENHEUER

K. O. KIEPENHEUER
(1910–1975)

TABLE OF CONTENTS

PREFACE

At the XV. General Assembly of the International Astronomical Union in Sydney 1973, Commission 10 for Solar Activity requested the incoming Organising Committee to establish a small group to recommend a standard nomenclature for solar features and to prepare an illustrated text which would clear the jungle of terms for the benefit of solar physicists as well as of theoreticians and research workers in related fields. The challenge was taken up by the president of Commission 10, Prof. K. O. Kiepenheuer, and his persuasive advocacy has led eventually to the present book.

In the course of the work, the declared aim but not the basic purpose was revised. Rather than prepare a list of standard terms, we have preferred to collect together all the terms that appear in current English-language literature. Synonyms and partially overlapping terms are all recorded for the most part without prejudice. Each has been defined as exactly as possible with the hope that in the future they may be used and understood without ambiguity. It would be a step on the road to standardisation if these terms were not re-used for new phenomena. New observations and new theories will lead to reappraisals and redefinitions so the Glossary is intended more as a guide to the present situation than as a rule-book.

More importantly, the Glossary is designed to be a technical dictionary that will provide solar workers of various specialties, students, other astronomers and theoreticians with concise information on the nature and properties of phenomena of solar and solar-terrestrial physics. A further section has the reciprocal aim of introducing to non-theorists terms that are most likely to be encountered in the description and interpretation of the observations. Each term, or group of related terms, is given a concise phenomenological and quantitative description, including the relationship to other phenomena and an interpretation in terms of physical processes. Of course, this has not been uniformly possible; a few things are well-understood, many are poorly understood. Despite the dangers of subjective judgement we have attempted to distinguish clearly the current status of each topic. We hope that the value of focussing attention on what is not known outweighs any injustice or irritation given to individual authorities. In the same spirit, the references have not been chosen for priority or topicality but for convenience, clarity and their overview. They are intended to lead the non-specialist reader into the literature. Books and journals not readily accessible have for the most part been avoided.

In such an undertaking as this, a good illustration is worth more than a thousand words. The pictures reproduced here are often more than illustrations, they are, in a very real sense, the phenomena themselves. We cannot approach them more closely or investigate them more directly. That is why they receive so much attention and why it is essential for the non-specialist to appreciate in what a wilderness the solar physicist must walk. Physics cannot as yet construct a priori a model that reproduces these pictures, so the attempt to understand these patterns, their evolution and relationships is still very fruitful and worthwhile. Starting from this premise the Glossary attempts to define these patterns and fix them into as wide a framework as possible so that the reader may appreciate their contribution to the overall scheme. Hopefully it will convince him that solar physics is concerned with the description of "species" solely to elucidate the inner workings of the basic physical processes that control our Sun, the archetypal star, and our environment.

Too many people have assisted and contributed to this volume for individual acknowledgement; they all have our warmest thanks. The editors would like to record their gratitude to the various contributors for their unfailing co-operation, patience and tolerance, and to record their debt to the late Prof. K. O. Kiepenheuer. This book is one memorial to his energy and determination.

The editors are aware that they have been unable to do justice to everybody for which they take full responsibility and offer a comprehensive apology. They are also aware that solar physics is a rapidly developing science and that the present picture will change. They welcome general criticism as well as further suggestions for inclusion, exclusion and revision.

A.B. and C.D.

LIST OF AUTHORS

Dr. J. M. Beckers (4), Sacramento Peak Observatory, Sunspot, N.M. 88349, U.S.A.

Dr. A. Bruzek, (7, 8, ed.), Fraunhofer Institut, Schöneckstr. 6, D-7800 Freiburg, F.R.G.

Dr. H. W. Dodson-Prince (9), McMath-Hulbert Observatory, 895 Lake Angelus Road North, Pontiac, Mich. 48055, U.S.A.

Dr. C. Durrant (1, 12, ed.), Fraunhofer Institut, Schöneckstr. 6, D-7800 Freiburg, F.R.G.

Dr. A. D. Fokker (11), Sterrekundig Instituut te Utrecht, Zonnenburg 2, Utrecht.

Dr. J. W. Harvey (3), Kitt Peak National Observatory, 950 North Cherry Avenue, P.O. Box 26732, Tucson, Ariz. 85726, U.S.A.

Dr. R. Howard (2), Hale Observatories, 813 Santa Barbara Street, Pasadena, Calif. 91101, U.S.A.

Dr. C. Jordan (5), Dept. of Theoretical Physics, University of Oxford, 12 Parks Road, Oxford OX1 3PQ.

Dr. S. Koutchmy (6), Institut d'Astrophysique, CNRS, 98 bis, Bd Arago, F-75014 Paris, France.

Mme. M. J. Martres (7), Observatoire de Paris, Section d'Astrophysique, F-92190 Meudon, France.

Prof. V. L. Patel (14), Dept. of Physics and Astronomy, University of Denver, University Park, Denver, Colo. 80208, U.S.A.

Prof. I. W. Roxburgh (1), Department of Applied Mathematics, Queen Mary College, University of London, Mile End Road, London E1 4NS.

Dr. L. Svalgaard (13), Solar Science Company, P.O. Box 7027, Palo Alto, Calif. 94305, U.S.A.

Dr. E. Tandberg-Hanssen (10), Astronomy and Solid State Physics Division, Space Science Laboratory, Marshall Space Flight Center, Ala. 35812, U.S.A.

(The numbers in brackets indicate the section of the Glossary prepared by the author.)

LIST OF FIGURES

1. SOLAR INTERIOR

C. J. DURRANT and I. W. ROXBURGH

1.1. Solar Standard Model

The interior model of the present Sun is obtained by taking one solar mass of plasma with the same chemical composition as the present outer layers supplemenied by a standard helium abundance and evolving it for 4.5×10^9 yr. No mixing of core material with the unevolved envelope is usually allowed. This may not be legitimate, but such uncertainties change the details, not the overall structure of the core.

Today, the model predicts an extended core since the p-p reaction, which releases almost all the luminous energy of the Sun, is only weakly temperature dependent. The extent of this energy source together with the temperature dependence of the bound-free

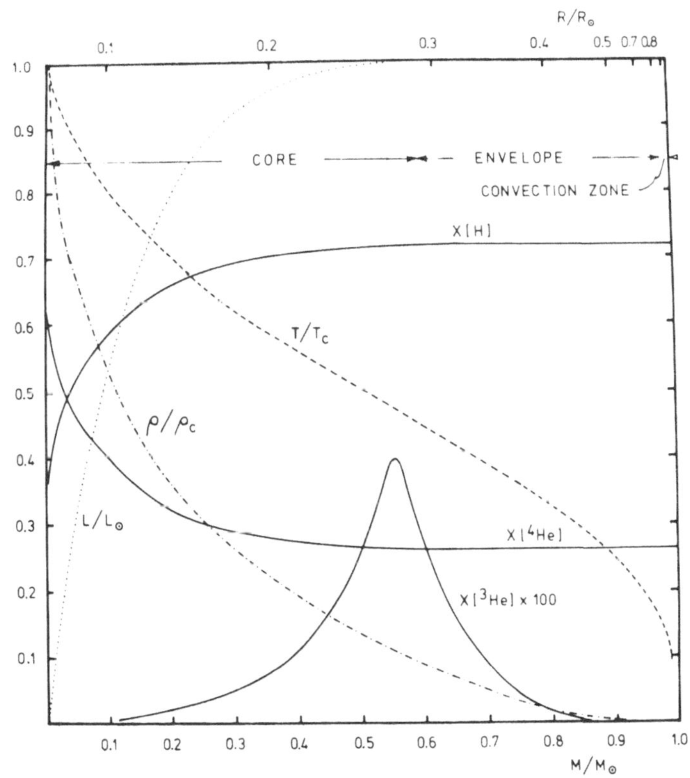

Fig. 1.1. Composite standard solar interior model. Plotted as a function of solar radius and mass are the temperature T, the density ρ, the luminosity L, and the fractional abundances by mass of H, ^3He and ^4He. Scaling parameters are $T_c = 1.6 \times 10^7$ K; $\rho_c = 1.6$ g cm^{-3}; $L_\odot = 3.86 \times 10^{33}$ erg s^{-1}; $R_\odot = 6.96 \times 10^{10}$ cm; $M_\odot = 1.99 \times 10^{33}$ g.

Bruzek and Durrant (eds.), Illustrated Glossary for Solar and Solar-Terrestrial Physics.
1–6.

and free-free opacities guarantee sub-adiabatic temperature gradients in the core, i.e. there is no convection.

In the outermost part of the envelope, however, a single convection zone occurs that is efficient in all but the outermost, just sub-photospheric, layers. It is these layers which determine the entropy of the adiabat in the efficient region and thus the model as a whole. Since a full theory of such inefficient convection is lacking, it leaves a free parameter in the model which must be determined by matching the calculated radius to the observed solar radius. Below the convection zone substantial convective overshoot may occur which would mix envelope material into nuclear burning (Li^7) regions.

See Figure 1.1.

1.2. Internal Rotation

In non-spherically symmetric fluids, such as the rotating Sun, it is not in general possible to satisfy both the radiative and hydrostatic equilibrium conditions. Both break down and the differential temperatures along level surfaces drive a meridional circulation. In stellar interiors these *Eddington-Sweet currents* are very slow (circulation time $\approx 10^{12}$ yr) and would not appreciably alter the angular momentum distribution in the Sun within its main-sequence lifetime. Furthermore, a molecular weight gradient, such as would build up at the boundary of the core, tends to choke the overall circulation as extra work is necessary to lift the denser material. Separate circulations then appear in core and envelope and no mixing occurs unless the core rotation is very rapid.

Dynamically driven flows would be much more important in redistributing angular momentum in the Sun. For barytropic liquids, in which the density is a function of pressure alone, variation of Coriolis force can be balanced only by viscous stresses which necessitate a rapid circulation and transport of angular momentum from rapidly spinning to slowly spinning regions. This is *spin-down* or *Ekman pumping.* Stars, however, are not barytropic and Coriolis terms can be balanced by variations of both density and temperature unless the shear is so great that a locally unstable density gradient arises. In the latter case, a dynamically driven mixing will occur but only in a thin boundary region. More effective in transporting angular momentum are secular instabilities which arise because radiative transfer smoothes out the temperature variations and removes the buoyancy force balancing the Coriolis force. These *Goldreich-Schubert-Fricke instabilities* lead, on a Kelvin-Helmholtz timescale of 10^7 yr, to a marginally stable angular velocity distribution which satisfies the stability criteria for the same system but composed of inviscid, incompressible fluid. The angular velocity must increase inwards but angular momentum will diffuse slowly outwards to the convection zone whence it is carried away by the solar wind. The timescale for the production of more or less uniform rotation is not known.

Uniform rotation would be achieved far more rapidly if even a weak magnetic field coupled the solar core to the outer envelope.

Recent solar oblateness measurements show no evidence of a rapidly rotating solar core.

Benton, E. R. and Clark, A.: 1974, *Ann. Rev. Fluid Mech.* **6**, 257.
Dicke, R. H.: 1972, *Astrophys. J.* **171**, 331.
Mestel, L.: 1965, *Stars and Stellar Systems* **8**, 465.

1.3. Energy Generation

As a main-sequence star, by far the most important energy source in the Sun is the thermonuclear conversion of four protons into one helium atom which releases the binding energy of 26.73 MeV.

At temperatures less than $\approx 2.3 \times 10^7$ K most energy is provided by the *proton-proton* (p-p) *chain*, a series of reactions that yields 26.20 MeV in gamma rays and 0.53 MeV in neutrinos.

$$p(p, e^+ + \nu)d$$
$$e^+(e^-)\gamma$$
$$d(p, \gamma)^3He$$
$$^3He(^3He, p + p)^4He.$$

At temperatures above 1.4×10^7 K alternative terminations involving heavier species are more frequent; firstly

$$^3He(^4He, \gamma)^7Be$$

then either $^7Be(e^-, \nu)^7Li$, $^7Li(p, \alpha)^4He$ (*pep reaction*)

or $^7Be(p, \gamma)^8B$, $^8B(,e^+ + \nu)^8Be^*$, $^8Be^* \rightarrow 2\,^4He$

The latter reaction produces neutrinos which should be detectable at the Earth.

At temperatures above 1.7×10^7 K a very different reaction catalysed by carbon, nitrogen and oxygen (hence *CNO cycle*) takes over. This is very sensitive to temperature (rate proportional to T^{17}) whereas the p-p chain is only weakly dependent (rate proportional to T^4). In solar models the CNO energy production is strongly concentrated at the core but even there provides only 6% of the total energy production rate in the standard model. A lower core temperature, that would be consistent with the observed neutrino flux, would produce even less energy by the CNO cycle.

Novotny, E.: 1973, *Introduction to Stellar Atmospheres and Interiors*, Oxford University Press, New York, p. 245.

1.4. Solar Neutrinos

The measurement of the solar neutrino flux is an important test of the solar standard model. The opacity of stellar material for neutrinos is so low that they escape directly from the solar core where they are produced in thermonuclear reactions. They should thus provide an unambiguous diagnostic for internal conditions. Davis has devised an experiment to detect the neutrinos from the side reactions

$$^7Be(e^-, \nu)Li^7$$
$$^8B(,e^+ + \nu)^8Be^*.$$

Most captures are expected from the latter.

Unfortunately the observations have given variable results and remain ambiguous. Standard models predict capture rates for the experiment of 6 SNU (1 SNU = 10^{-36} captures per target particle per second) whilst rates of only 1–3 SNU have been measured.

Many attempts have been made to account for the discrepancy, a common suggestion being that the core temperature is lower than that predicted by the standard model. Since the neutrino-producing reactions are highly temperature sensitive whilst the luminosity producing reactions (p-p chain) are only weakly so, it is possible, by mixing, to have at the present time a less evolved Sun with the same luminosity but lower neutrino flux.

An attractive model – the *solar spoon* – producing such mixing invokes a nuclear instability due to the very high temperature dependence of the conversion of He^3 to He^4 at the edge of the core where unburnt He^3 tends to accumulate. The energy release drives a slow mixing across the core boundary.

Roxburgh, I. W.: 1976, in V. Bumba and J. Kleczek (eds.), 'Basic Mechanisms of Solar Activity', *IAU Symp.* **71**, 453.
Trimble, V. and Reines, F.: 1973, *Rev. Mod. Phys.* **45**, 1.

1.5. Convection Zone

A radially symmetric stellar envelope can remain static with the radiation absorbed everywhere being exactly balanced by that emitted (radiative equilibrium) until the local temperature gradient required to maintain the constant heat flux becomes marginally superadiabatic. Because of the large Rayleigh numbers in stellar interiors this Schwarzschild criterion for instability suffices. In the solar envelope the onset of hydrogen recombination causes the value of the adiabatic gradient to fall and convection to set in. In these dense regions the convection is efficient, i.e. all the thermal energy may be considered to be transported by the convective motions (convective equilibrium) and the superadiabaticity required is so small that the temperature gradient may be taken to be the adiabatic gradient. Whilst the pressure increases inwards through this region the ionization holds down the temperature. This forces steep temperature gradients to appear below the ionization zone and convection to extend for a considerable distance into this region.

At the base of the solar photosphere the increasing H^- opacity requires a radiative temperature gradient considerably above the adiabatic. In this case, however, the density is low so the resulting convection is inefficient; neither all the energy is transported by convection, nor is the temperature gradient close to the adiabatic. Some convective theory is necessary to treat this regime.

See Figure 1.2.

Cox, J. P. and Giuli, R. T.: 1968, *Principles of Stellar Structure*, Gordon and Breach, New York.
Spruit, H. C.: 1974, *Solar Phys.* **34**, 277.

1.6. Convective Theory

The theory of convection tries to describe the steady state established in convectively unstable non-linear systems. Its complexity forces a choice between a rigorous treatment of a simple system or a schematic treatment of a more complex model. Much important progress has been made recently in the former area but usually these treatments impose the Boussinesq assumption, assume an isolated layer and require a high Prandtl number (the ratio of viscosity to conductivity). Since none of these requirements are met in the Sun, the results must be applied with caution.

Resort is often made to heuristic descriptions of turbulent convection. One such description, the *mixing-length theory*, assumes that the convective transport processes are analogous to molecular processes and can be described in terms of a single mixing-length corresponding to the molecular mean-free path. This length is a free parameter usually chosen to be comparable to the local density scale height. It should be stressed that this is still a Boussinesq theory.

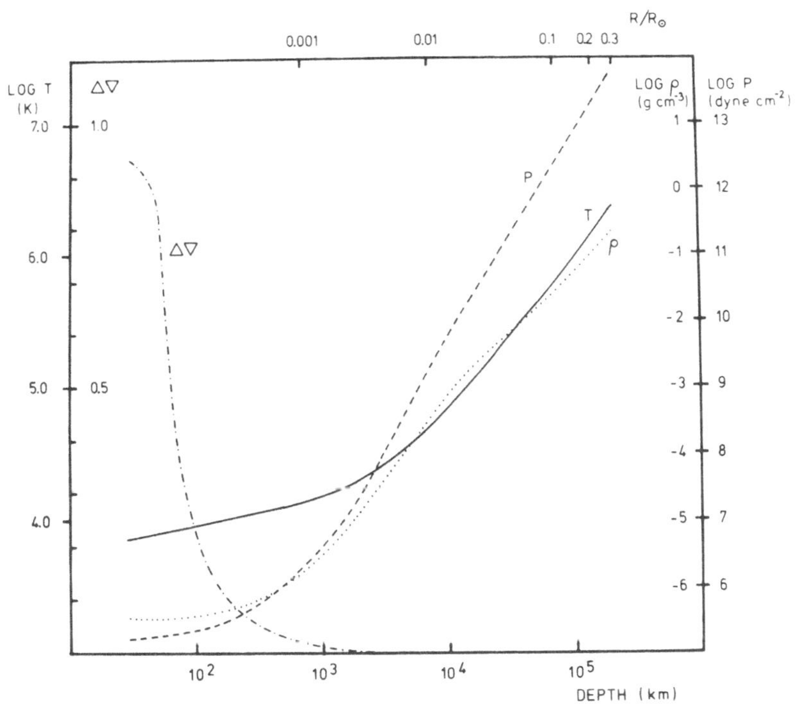

Fig. 1.2. Mixing-length model of the solar convection zone after Spruit (1974). Temperature T, density ρ, gas pressure P, and superadiabatic gradient $\Delta\nabla$, are plotted as a function of depth below the base of the photosphere.

On the other hand, it has been argued that in convection extending over several (slowly varying) scale heights large-scale cells will be preferred energetically to eddies comparable in size to the local scale height. The solar convection zone can then be divided into three regimes giving rise to cells of different sizes manifested in the granulation, supergranulation and giant cells.

Simon, G. W. and Weiss, N. O.: 1968. *Z. Astrophys.* **69**, 435.
Spiegel, E. A.: 1971, *Ann. Rev. Astron. Astrophys.* **9**, 323.

1.7. Boussinesq Assumption

This is a simplifying assumption usual in non-linear treatments of convection in compressible media. It requires the vertical extent of the medium to be much less than the density and pressure scale heights. Compressibility (density fluctuations) then give rise only to buoyancy forces and not to inertial forces. The term is sometimes taken to imply also the *anelastic assumption.* This requires all velocities to be much less than that of sound so that all pressure waves are filtered out.

Turner, J. S.: 1973, *Buoyancy Effects in Fluids*, Cambridge, p. 9.

2. SOLAR CYCLE, SOLAR ROTATION AND LARGE-SCALE CIRCULATION

R. HOWARD

2.1. Solar (Activity) Cycle

The periodic variation in intensity or number of the various manifestations of solar activity is referred to as the solar activity cycle. Examples of features or indices that show such variations are: Wolf Number (*Sunspot cycle*), plages, prominences, 2800 MHz radio emission and flares. The period of this cyclic variation is approximately 11 yr (*11-yr cycle*), although in this century the average period has been closer to 10 yr.

The rise to maximum in the Wolf Number and the other indices is typically shorter than the decline from maximum to minimum. Figure 2.1 shows the variation of the Wolf Number for the last three solar cycles.

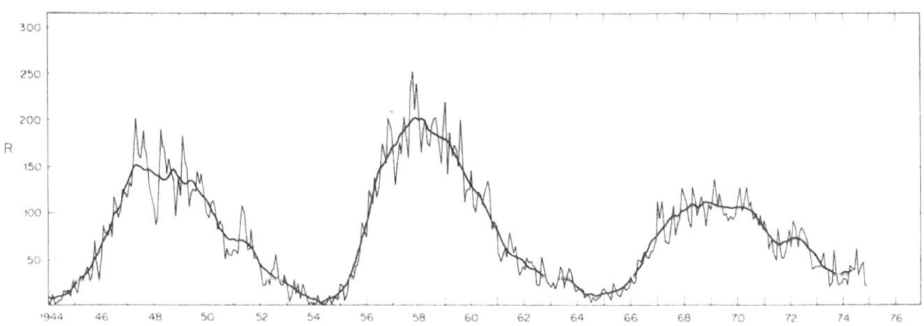

Fig. 2.1. Plot of the monthly average Wolf Number (R) for the past three solar cycles. The smooth curve is the smoothed mean

$$\bar{R}_0 = \frac{R_{-6} + R_{+6} + 2\sum_{-5}^{5} R_i}{24}.$$

The first sunspots of a new cycle are formed at relatively high latitudes. The mean latitude of spot formation decreases during the cycle until the spots are near the equator at the end of the cycle.

This variation of sunspot latitude with phase of the solar cycle has been given the name *Spörer's Law*, after the first person who investigated the effect in detail, The *butterfly diagram* is a graphical representation of the variation of the latitude dependence of sunspot occurrence. It was first plotted by Maunder in 1922. There is an overlap of approximately 3 yr near the minimum of activity between the lower-latitude activity of the old cycle and the high-latitude activity of the new cycle (Figure 2.2).

Kiepenheuer, K. O.: 1953, in G. P. Kuiper (ed.), *The Sun*, Univ. of Chicago Press, Chicago, p. 322.
Waldmeier, M.: 1955, *Ergebnisse und Probleme der Sonnenforschung*, Akad. Verlagsges. Geest and Portig, Leipzig, pp. 139–160.

Bruzek and Durrant (eds.), Illustrated Glossary for Solar and Solar-Terrestrial Physics.
7–12.

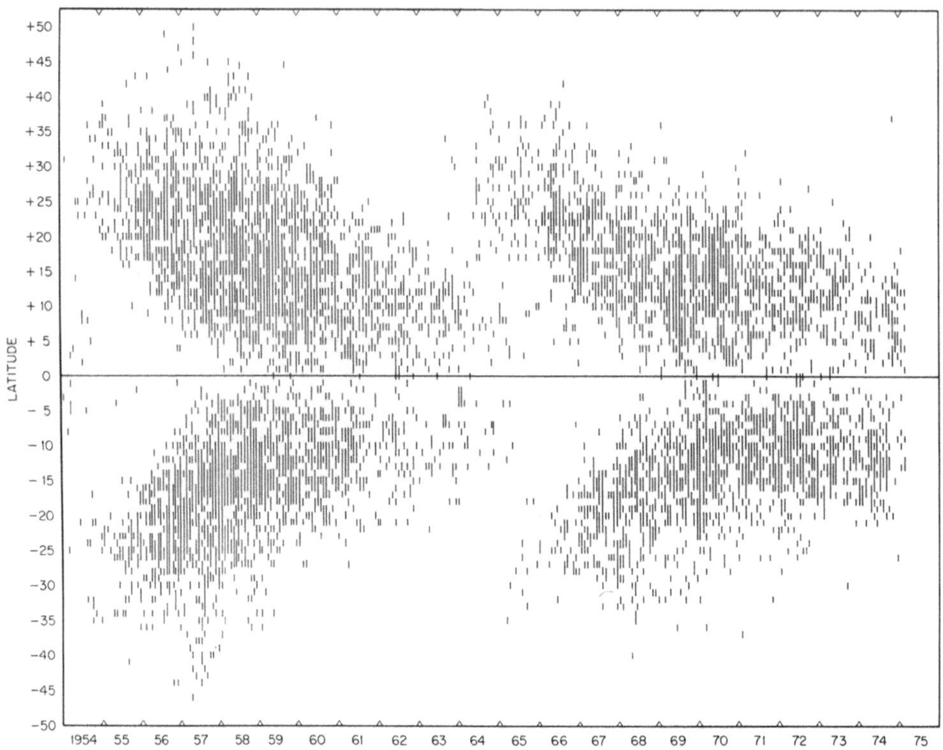

Fig. 2.2. Butterfly Diagram constructed from Mt. Wilson sunspot observations. A vertical line covering one degree in latitude is plotted for each sunspot group centred at that latitude within a Carrington rotation. This diagram shows the last two solar cycles.

2.2. Hale Cycle

The magnetic polarities of preceding and following sunspots in each hemisphere reverse from one activity cycle to the next. In Solar Cycle 20 which began in October, 1964, the preceding spot polarities on the northern hemisphere were predominantly of negative (or south) polarity (magnetic vector pointing into the Sun). The following polarities were positive (north). On the southern hemisphere the preceding polarities were positive (north) and the following polarities were negative (south). In the next cycle these polarities in each hemisphere will be reversed. Thus the magnetic activity exhibits a 22-yr repeatability. This is referred to as the Hale Cycle.

Hale, G. E. and Nicholson, S. B.: 1938, Publ. Carnegie Inst., No. 498.

2.3. Long-Term Activity Variations

There is some evidence for a modulation of the amplitude of the 11-yr cycle in a period of about 80 yr. Although sunspot numbers may be traced back with confidence only about 2½ centuries, auroral records have been used to trace the level of activity back more than 2 millenia. Other long periods that have been reported are 200, 400, and 600 yr.

Cole suggests that the free-running length of the cycle is 11.8 yr but that it is triggered every 10.45 yr giving a phase modulation period of 190 yr. The amplitude is modulated by a second periodicity of 11.9 yr.

Eddy has pointed out that for an interval of about 70 yr, starting in 1645, very few sunspots were observed, and little or no evidence for a sunspot cycle may be found. Other records, such as aurorae, naked-eye spots, and the C^{14} time scale calibration tend to support the existence of a low level of solar activity during this interval which is referred to as the *Maunder Minimum.*

Bonov, A. D.: 1973, in J. Xanthakis (ed.), *Solar Activity and Related Interplanetary and Terrestrial Phenomena*, Springer, Berlin, p. 83.
Cole, T. W.: 1973, *Solar Phys.* **30**, 103.
Eddy, J. A.: 1976, *Science* **192**, 1189.
Henkel, R.: 1972, *Solar Phys.* **25**, 498.

2.4. Dynamos

A MHD dynamo requires fluid motions in a conducting medium to create currents that generate magnetic flux. Kinematic dynamos, in which the flows are prescribed without consideration of any build-up of magnetic forces, have been most commonly studied and it has been found that a dynamo will result from almost any motion of sufficient vigour and complexity.

Most solar dynamo models use the differential rotation of the convection zone to stretch a large-scale poloidal (in planes through the rotation axis) magnetic field into a large-scale toroidal (wrapped around the rotation axis) field. This toroidal field then generates new poloidal field by twisting in small-scale convective elements due to Coriolis forces. The asymmetry between rising and falling elements in a stratified medium ensures that these cyclonic motions have a net large-scale effect.

Mean-field theories use the differential rotation (the ω-effect) and a model of the convective motions (the α-effect) to follow the growth of the large-scale field. A cyclic α-ω dynamo model is obtained by balancing this generation against dissipation of magnetic flux. Turbulent diffusion is used to convert the large-scale into intermediate- and small-scale flux. Some of the former is lost directly by diffusion through the solar surface and the rest by ohmic dissipation of the small-scale flux. The α-effect and turbulent diffusivity can be estimated using mixing-length convective models, but the form of differential rotation remains a free parameter. Suitable choices give models that reproduce many global features of the solar activity cycle.

Solar hydromagnetic dynamo models, in which the field and flow equations are satisfied self-consistently have not yet been developed.

The original semi-empirical *models of Babcock* and *Leighton* were very similar except Babcock appealed to magnetic buoyancy to raise his toroidal loops and Leighton used the granular/supergranular motions to provide the horizontal diffusive flux transport. Indeed the realisation that the diffusion may be inhomogeneous and anisotropic is probably significant.

Babcock, H. W.: 1961, *Astrophys. J.* **133**, 572.

Leighton, R. B.: 1964, *Astrophys. J.* **140**, 1547.
Stix, M.: 1974, *Astron. Astrophys.* **37**, 121.

2.5. Differential Rotation

The Sun does not rotate as a solid body; low latitudes rotate at a faster angular rate than do high latitudes. This effect is called differential rotation. Its magnitude is determined either by timing tracers – sunspots, faculae, filaments, etc. – as they cross the solar disk or by the use of the Doppler effect in spectral lines. Each method yields a different latitude dependence though most tracers show the same rotation rate at the equator. The difference between the tracer and the Doppler rates has been attributed to a rotation rate that increases with depth below the visible surface.

There is no generally accepted explanation for the existence of differential rotation in the Sun. See Figure 2.3.

Gilman, P. A.: 1974, *Ann. Rev. Astron. Astrophys.* **12**, 47.

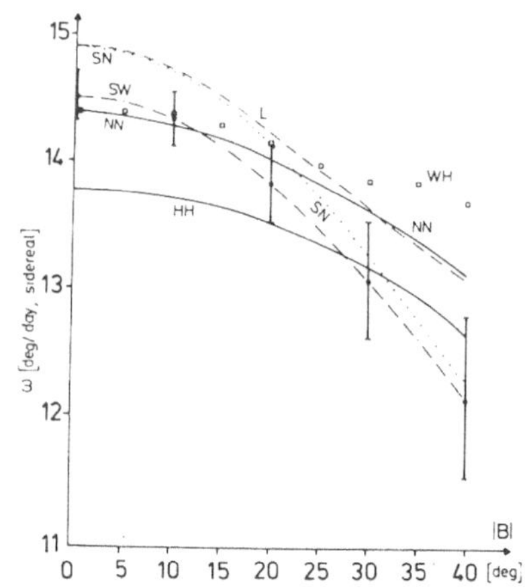

Fig. 2.3. Differential rotation. Rotation rates for sunspots (NN = Newton and Nunn, 1951), the photospheric plasma (HH = Howard and Harvey, 1970), chromospheric plasma in Hα (L = Livingston, 1969), photospheric magnetic field (WH = Wilcox and Howard, 1970), the coronal plasma (SN = Simon and Noyes, 1972), the Ca II K mottles (SW = Schröter and Wöhl, 1975) (after: Schröter, E. H. and Wöhl, H.: 1975, *Solar Phys.* **42**, 3).

2.6. Rotation of the Convection Zone

The differential rotation is considered to be a consequence of the interaction of rotation and convection, although the details of the interaction can only be calculated for approximate phenomenological models of solar convection since there is, as yet, no theory of compressible turbulence. The models fall into two classes; the first class assumes that the dynamics of the rotating solar convective zone are similar to those of a rotating shell of liquid heated from below. The resulting convective motions redistribute the angular momentum until, in a steady state, the rotation is non-uniform. The second class is based on attempts to model the effect of turbulence by 'eddy transport coefficients' which are anisotropic or inhomogeneous, or both. In this case, a large-scale circulation cell redistributes angular momentum which again results in differential rotation.

The one firm conclusion from these analyses is that any kind of motion in a rotating fluid will lead to equatorial acceleration or deceleration. All models have unknown adjustable parameters which by judicious choice can produce agreement between the model and the observed differential rotation, but they almost all predict an embarrassingly large and unobserved angular variation of heat flux from the Sun. However, a detailed model including compressibility may lead to a solution of the flux problem.

Durney, B. R., Gilman, P. A., and Stix, M.: 1976, in V. Bumba and J. Kleczek (eds.), 'Basic Mechanisms of Solar Activity', *IAU Symp.* **71**, 479.

2.7. Carrington Rotation

More than a century ago Carrington determined an average synodic rotation rate for sunspots of 27.2753 days (sidereal rate 25.38 days). Rotations are numbered starting with No. 1 on 9 November 1853. Rotation No. 1636 commenced at 15.95 December 1975.

This rotation defines a longitude grid; each rotation starts at the instant at which 0° long. crosses the central meridian. The longitudes increase from 0° to 360° going from east to west on the Sun.

Carrington, R. C.: 1863, *Observations of the Spots on the Sun*, Williams and Norgate, London, p. 16.

2.8. Oblateness (Solar)

If the observed surface rotation rate of the Sun were to extend more or less unchanged into the deep interior, a slight oblateness of the disk, $\Delta r/r = 10^{-5}$, or about 0.01″ in the radius, would be expected. Some recent measurements by Dicke have indicated an oblateness of about five times this value.

Dicke's explanation requires a rapidly rotating core (period < 2 days) to distort the gravitational potential at the surface. If the contribution of this rotation to the precession of the perihelion of Mercury is removed, the remaining relativistic correction would favour the scalar-tensor theory of gravitation over Einstein's.

The results of later observations have cast doubt on the validity of the original measurements.

Dicke, R. H. and Goldenberg, H. M.: 1974, *Astrophys. J. Suppl.* **27**, 131.
Hill, H. A. *et al.*: 1974, *Phys. Rev. Letters* **33**, 1497.

2.9. Meridional Flow

Some models of the differential rotation of the Sun require an axisymmetric meridional (in planes through the rotation axis) flow equatorward in order to transport angular momentum. Some observations tend to support such a large-scale motion whilst others show no evidence of it. A pole-equator temperature difference which could drive such a flow appears not to exist.

Gilman, P. A.: 1974, *Ann. Rev. Astron. Astrophys.* **12**, 47.
Howard, R.: 1971, *Solar Phys.* **16**, 21.
Plaskett, H. H.: 1966, *Monthly Notices Roy. Astron. Soc.* **131**, 407.

2.10. Giant Cells or Large-Scale Circulation

There is some evidence in the Doppler shifts of spectral lines for the existence of large-scale circulation patterns. In addition, the distribution across the solar surface of active regions and weak magnetic fields has been suggested to result from a large-scale cellular pattern. Such cells have a typical dimension of 300 000 km and a typical surface velocity ≈ 0.1 km s^{-1}.

Alternative explanations have been offered, notably in terms of Rossby waves and in terms of rotationally-distorted giant convection cells.

Bumba, V. and Howard, R.: 1965, *Astrophys. J.* **141**, 1502.
Gilman, P. A.: 1974, *Ann. Rev. Astron. Astrophys.* **12**, 47.
Simon, G. W. and Weiss, N. O.: 1968, *Z. Astrophys.* **69**, 435.

3. NON-SPOT MAGNETIC FIELDS

J. W. HARVEY

3.1. General Magnetic Field

The notion that the Sun should have a permanent dipolar magnetic field arose in the last century. Modern observations cast doubt on the existence of such a field. Today, the term general magnetic field is often used interchangeably with polar field. The problem of determining the existence of a true general magnetic field is that of eliminating the effect of transient magnetic fields which collect near the poles. See polar field and large scale magnetic field.

Bigelow, F. H.: 1889, *The Solar Corona*, Smithsonian Institution.
Hale, G. E.: 1923, *Astrophys. J.* **38**, 27.
Piddington, J. H.: 1976, *Astrophys. Space Sci.* **47**, 319.
Severny, A. B..: 1971, in R. Howard (ed.), 'Solar Magnetic Fields', *IAU Symp.* **43**, 675.
Stenflo, J. O.: 1971, in R. Howard (ed.), 'Solar Magnetic Fields', *IAU Symp.* **43**, 714.

3.2. Polar Magnetic Field

Observations of the Sun's polar regions (above 60° lat.) usually reveal a well-defined pattern of magnetic fields, the polar field, whose gross characteristics vary slowly with a time-scale of years. It is not to be confused with the hypothetical general magnetic field. There is little or no evidence for a general, smoothly distributed magnetic field in the polar regions. The polar field appears to be composed of small magnetic elements, as elsewhere on the Sun, which individually change on a short time-scale but whose total flux for the entire polar region is $\approx 10^{21}$ Mx. The polarity averaged over the cap area is not always well defined but generally shows opposite polarities in the two hemispheres (Figure 3.1), although it is not infrequent that both hemispheres show the same polarity for some weeks or months. The polarities of the polar fields have been observed to reverse at or following the time of activity maximum.

According to present ideas the polar field is largely composed of the remnants of the trailing parts of active regions formed at lower latitudes.

Gillespie, B. *et al.*: 1973, *Astrophys. J.* **186**, L85.
Howard, R.: 1974, *Solar Phys.* **38**, 283.
Pope, T. and Mosher, J.: 1975, *Solar Phys.* **44**, 3.
Severny, A. B.: 1971, in R. Howard (ed.), 'Solar Magnetic Fields', *IAU Symp.* **43**, 675.
Stenflo, J. O.: 1970, *Solar Phys.* **13**, 42.

3.3. Large-Scale Magnetic Field

Although the photospheric magnetic field is composed of small magnetic structures, these structures are organized into large-scale patterns as revealed by visual inspection and

Bruzek and Durrant (eds.), Illustrated Glossary for Solar and Solar-Terrestrial Physics.
13–20.

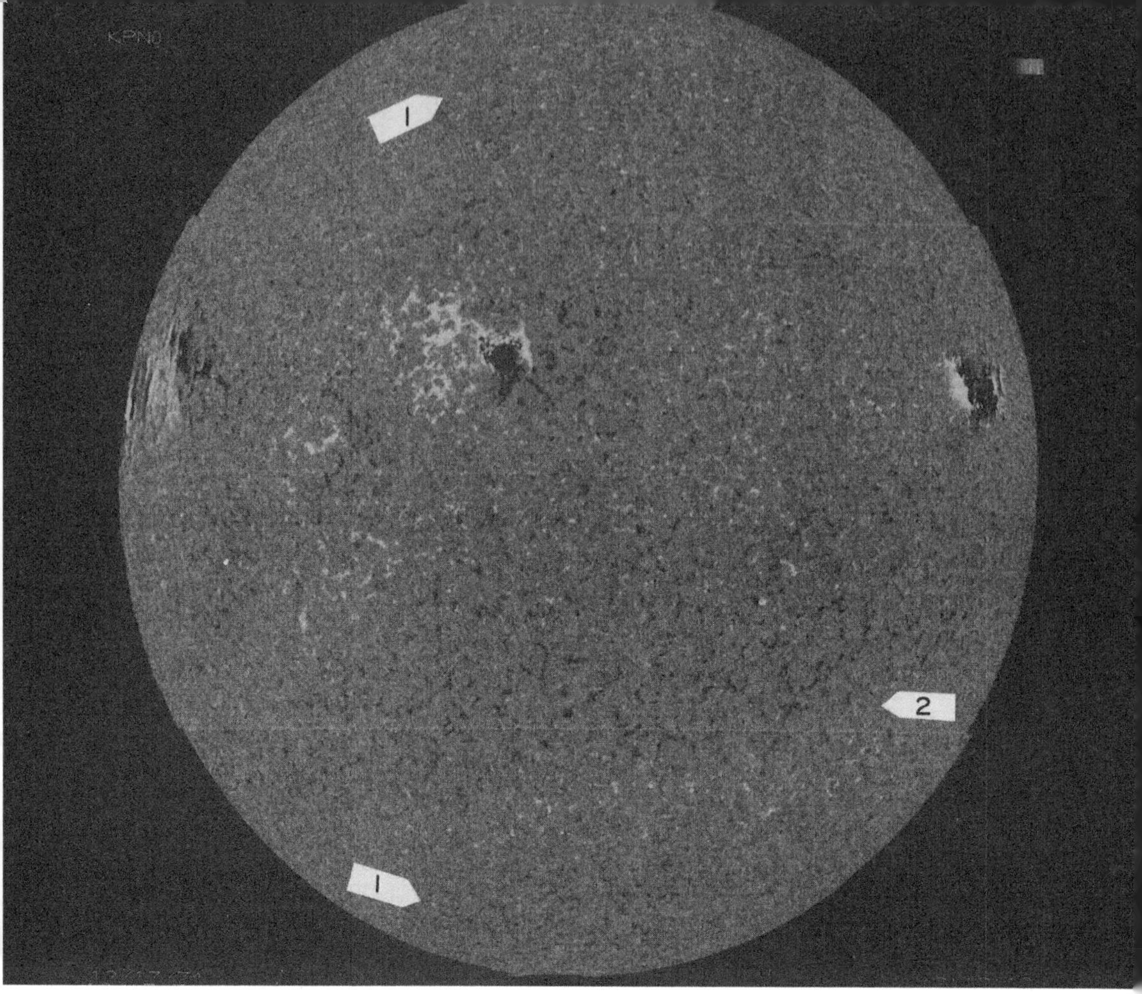

Fig. 3.1. Kitt Peak magnetogram 17 December 1974 showing (1) polar fields positive (bright) at the north and negative (dark) at the south and (2) a large-scale field pattern in the south. Note the network pattern in the magnetic fields. In the northern hemisphere are several active regions.

harmonic analysis of magnetograms such as in Figure 3.1. Active regions represent a medium-size scale. On a large scale the magnetic field usually exhibits an apparent dipole field at the poles and large areas predominantly having single polarities at lower latitudes. These large-scale patterns persist for long periods of time and partake of a rotation different from that of the non-magnetic photosphere. Large-scale patterns on the surface are manifest by actual physical connections in the corona covering immense distances.

Altschuler, M. D., Trotter, D. E., Newkirk, G., Jr., and Howard, R.: 1974, *Solar Phys.* **39**, 3.
Nakagawa, Y. and Levine, R. H.: 1974, *Astrophys. J.* **190**, 441.
Sheeley, N. R., Jr., Bohlin, J. D., Brueckner, G. E., Purcell, J. D., Scherrer, V., and Tousey, R.: 1975, *Solar Phys.* **40**, 103.
Wilcox, J. M. and Howard, R.: 1970, *Solar Phys.* **13**, 251.

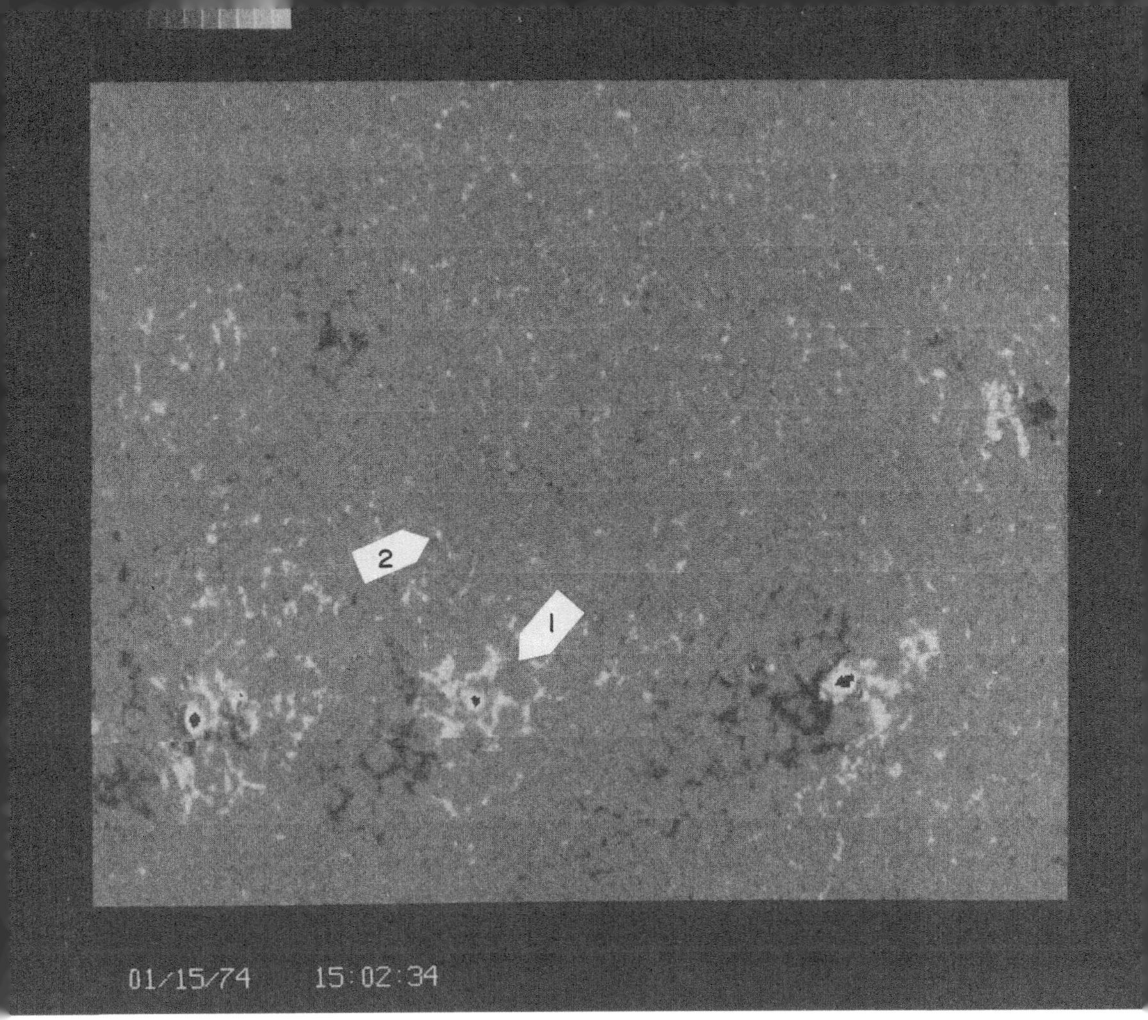

Fig. 3.2. Kitt Peak magnetogram 15 January 1974 showing (1) a magnetic puka, (2) an ephemeral region and a number of bipolar active regions. The area covered is 1000 by 1250″.

3.4. Sector Structure

The interplanetary magnetic field in the ecliptic can usually be divided azimuthally into a small number of sectors within which the magnetic field is predominantly directed toward or away from the Sun. These patterns can be traced back to the surface of the Sun where they correspond to large-scale weak fields directed inward or outward. The west side of a sector boundary tends to be quiet and the east side tends to be active. Magnetic arcades cross the sector boundary.

Altschuler, M. D., Trotter, D. E., Newkirk, G., Jr., and Howard, R.: 1974, *Solar Phys.* **39**, 3.
Schatten, K. H.: 1972, in C. P. Sonnett *et al.*, (eds.), *Solar Wind*, NASA SP-308, p. 65.
Svalgaard, L. and Wilcox, J. M.: 1975, *Solar Phys.* **41**, 461.
Wilcox, J. M.: 1971, in R. Howard (ed.), 'Solar Magnetic Fields', *IAU Smp.* **43**, 744.
Wilcox, J. M. and Svalgaard, L.: 1974, *Solar Phys.* **34**, 461.

3.5. Unipolar Magnetic Region

A large-scale photospheric region where the magnetic elements are predominantly of one polarity. Examples are polar regions most of the time and the trailing parts of major active regions several months after their disintegration.

Babcock, H. W. and Babcock, H. D.: 1955, *Astrophys. J.* **121**, 349.
Bumba, V. and Howard, R.: 1965, *Astrophys. J.* **141**, 1502.

3.6. Magnetic Puka

Puka is a Hawaiian word that means hole or tunnel. A magnetic puka refers to a region remaining free of significant amounts of magnetic flux for 4–7 days even though completely surrounded by flux. Figure 3.2 shows a puka.

Livingston, W. C. and Orrall, F. Q.: 1974, *Solar Phys.* **39**, 301.

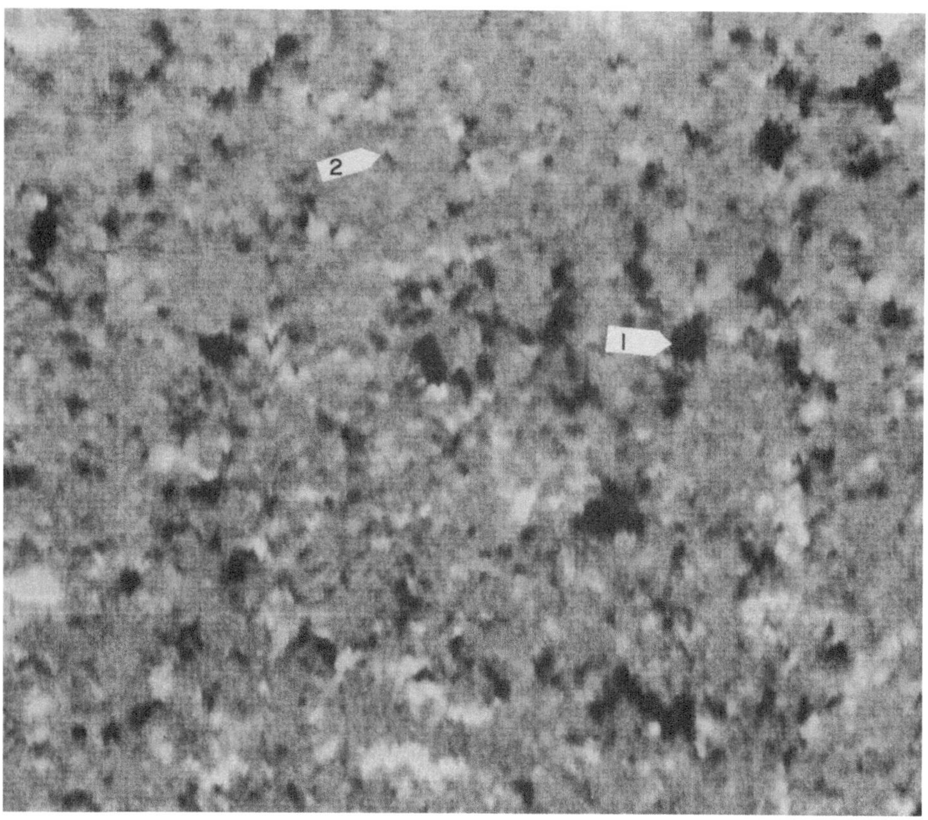

Fig. 3.3. Kitt Peak magnetogram at the disk centre in a quiet region 1 October 1974 showing resolution of non-spot magnetic fields into magnetic elements or fluxules. Also shown are (1) network and (2) inner network fields. The area covered is 225 by 260″ and the weakest features contain 5×10^{16} Mx.

3.7. Network Field

In the quiet photosphere most of the observable magnetic flux is concentrated in a network pattern usually assumed to follow the boundaries of supergranule cells. Such magnetic fields have been termed network fields. Magnetic flux can also be observed away from the network giving rise to the term *inner-network field*. Figure 3.3 shows both types of fields. The inner-network fields are characterized by mixed polarity, fluxes $\approx 10^{17}$ Mx per resolved element and lifetimes $\approx$30 min.

Livingston, W. C. and Harvey, J.: 1975, *Bull. Am. Astron. Soc.* 7, 346.

3.8. Magnetic Hills

On contour maps of magnetic field observations with moderate resolution, concentrations of flux appear as do hills on geographic elevation contour maps. Thus the term hill refers to a medium-scale concentration of magnetic flux.

Bappu, M. K. V., Grigorjev, V. M., and Stepanov, V. E.: 1968, *Solar Phys.* **4**, 409.

3.9. Magnetic Element or Fluxule

High spatial resolution observations of photospheric magnetic fields show that outside sunspots all fields are highly fragmented. These fragments have been called magnetic elements or, with reference to magnetographs which record magnetic flux rather than magnetic field strength, fluxules (Figure 3.3).

The elements tend to clump together with varying degrees of compactness to form larger magnetic features which in turn are organised into still larger-scale complexes – the network and active regions. The sizes of the smallest magnetic elements are almost certainly too small to be resolved. The assumed true size then influences both the total flux and mean magnetic field strength of the element. Some analyses suggest a characteristic size of 100–300 km, a flux of 3×10^{17} Mx and a field strength of 1–2 kG. This interpretation also requires 90% of all the surface flux to exist in this form. The lifetimes of the elements has not been well determined but is probably shorter than that of the larger aggregates such as knots.

It is surmised but not yet demonstrated that these magnetic elements coincide with facular points and filigree.

Harvey, J.: 1977, *Highlights of Astronomy*, **4**, in press.
Stenflo, J. O.: 1973, *Solar Phys.* **32**, 41.
Vrabec, D.: 1974, in R. G. Athay (ed.), 'Chromospheric Fine Structure', *IAU Symp.* **54**, 201.

3.10. Magnetic Rope (Flux Rope)

One old concept of the physical structure of magnetic fields, particularly near the photosphere, is that the field is organized in a hierarchy of twisted filaments, from threads through strands to large-scale bundles or ropes. Observational evidence is generally consistent with this idea. If the magnetic rope concept is not essentially correct then processes such as convection operate in such a way as to simulate the appearance of threads and ropes.

Babcock, H. W.: 1961, *Astrophys. J.* **133**, 572.
Jensen, E.: 1955, *Ann. Astrophys.* **18**, 127.
Parker, E. N.: 1955, *Astrophys. J.* **121**, 491.
Parker, E. N.: 1974, *Astrophys. J.* **191**, 245.
Piddington, J. H.: 1976, *Astrophys. Space Sci.* **40**, 73.

3.11. Magnetic Filament

This term conveys the idea that solar magnetic fields are thread-like in structure if they could be seen in three dimensions. In practice only a two-dimensional visualization is practical so the term filament as generally used is nearly synonymous with element, knot or fluxule. See magnetic rope.

Stenflo, J. O.: 1971, in R. Howard (ed.), 'Solar Magnetic Fields', *IAU Symp.* **43**, 101.

3.12. Magnetic Microturbulence

If magnetic field fluctuations occur on a spatial scale small compared with the scale of line formation then significant modification of emergent polarization may occur analogous to the differing effects of macro- and microturbulence on ordinary line formation. So far there is no certain evidence that very small-scale magnetic fluctuations exist.

Howard, R. and Bhatnagar, A.: 1969, *Solar Phys.* **10**, 245.
Staude, J.: 1970, *Solar Phys.* **12**, 84.
Stenflo, J. O.: 1977, in R.-M. Bonnet and P. Delache (eds.), 'The Energy Balance and Hydrodynamics of the Solar Chromosphere and Corona', *IAU Colloquium*, **36**, 143.
Unno, W.: 1959, *Astrophys. J.* **129**, 375.

3.13. Crossover Effect

This effect was first observed by Babcock in magnetic stars and refers to peculiar Zeeman patterns observed when the net polarity of a magnetic variable star crosses over from one polarity to the other. The explanation is that two oppositely directed, adjacent magnetic fields are observed without adequate spatial resolution so that a relative Doppler shift in the two regions with a magnitude similar to the Zeeman splitting produces a net Zeeman pattern with the observed anamolous polarizations. On the Sun the effect is fairly rare, having escaped notice until recently. What is observed is an apparently normal Zeeman triplet except the σ components have the same sense of circular polarization and the π component is oppositely circularly polarized.

Babcock, H. W.: 1951, *Astrophys. J.* **114**, 1.
Babcock, H. W.: 1956, *Astrophys. J.* **124**, 489.
Golovko, A. A.: 1974, *Solar Phys.* **37**, 113.
Grigorjev, V. M. and Katz, J. M.: 1972, *Solar Phys.* **22**, 119.
Moe, O. K.: 1968, in K. O. Kiepenheuer (ed.), 'Structure and Development of Solar Active Regions', *IAU Symp.* **35**, 202.

3.14. Magnetograph

Broadly, a magnetograph is any instrument or technique for mapping solar magnetic fields; in practice, it usually measures the polarization produced in spectrum lines by the Zeeman effect. Magnetographs are thus polarimeters though the complete Stokes vector is not necessarily measured.

Spectral isolation is provided by spectrographs, spectroheliographs, narrow-band and, in a few applications, broad-band filters. Detectors used include photomultiplier tubes, discrete and monolithic diode arrays, photographic plates and films and television camera tubes.

Almost any technique that can be imagined for measuring polarization has been proposed or used as a magnetograph. Most commonly, the polarization signal is detected through modulation of the light beam by optical retarders such as rotating quarter-wave or half-wave plates, or by electro-optical modulators (KDP crystals or Kerr cells). Instrumental polarization can be almost completely eliminated by suitable differential measurements.

A motor-driven tilted glass plate (*Doppler compensator*) is often used to centre the line profile automatically. It is also employed for the sensitive measurement of line shifts.

Currently the most routinely used instruments are the Babcock-type magnetograph, the video-magnetograph and the photographic spectrograph with a polarization analyser, which all measure just the longitudinal magnetic field; and the vector magnetograph for measuring the total field vector.

It is probable that magnetographs fail to resolve the smallest elements of non-spot fields. Quoted values give the field strength averaged over a few arcsec and can greatly underestimate the true field strengths.

Beckers, J. M.: 1971, in R. Howard (ed.), 'Solar Magnetic Fields', *IAU Symp.* **43**, 3.
Schröter, E. H.: 1973, *Mitt. Astron. Ges.* **32**, 55.
Vrabec, D.: 1974, in R. G. Athay (ed.), 'Chromospheric Fine Structure', *IAU Symp.* **56**, 224.

3.15. Stokesmeter

A Stokesmeter is a type of polarimeter or vector magnetograph which measures the complete state of polarization across one or more spectral lines. The light is analysed into four components producing four photographic spectra or four modulated photoelectric signals in order to determine the four Stokes parameters (the Stokes vector).

Ideally one would like the Stokes vector as a function of wavelength to allow analysis of observations in terms of complete line transfer models for homogeneous and inhomogeneous model atmospheres.

Baur, T. G., Curtis, G. W., Hull, H., and Rush, J.: 1974, in T. Gehrels (ed.), 'Planets, Stars and Nebulae Studied with Photopolarimetry', *IAU Colloquium* **23**, 246.
Cacciani, A. and Fofi, M.: 1971, *Solar Phys.* **19**, 270.
Harvey, J., Livingston, W., and Slaughter, C.: 1972, in *Proc. Conf. on Line Formation in the Presence of Magnetic Fields*, HAO-NCAR, Boulder, p. 227.
House, L. L., Baur, T. G., and Hull, H. K.: 1975, *Solar Phys.* **45**, 495.
Nishi, K. and Makita, M.: 1973, *Publ. Astron. Soc. Japan* **25**, 51.
Orrall, F. Q.: 1971, in R. Howard (ed.), 'Solar Magnetic Fields', *IAU Symp.* **43**, 30.
Wiehr, E.: 1974, *Solar Phys.* **35**, 351.
Wittmann, A.: 1973, *Solar Phys.* **33**, 107.

3.16. Lambdameter or Recording Doppler Comparator

A device for recording the wavelength variations of a spectrum line or lines in the direction perpendicular to dispersion. It is used on photographic spectra exclusively. The principle of operation is to determine the wavelength shift necessary to balance the measured degree of exposure in the two wings of a spectrum line as viewed through two slit apertures. It operates in the same manner as the Doppler compensator of a Babcock-type magnetograph.

Charvin, P., Rayrole, J., and Semel, M.: 1962, *Compt. Rend. Acad. Sci. Paris* **254**, 2289.
Evans, J. W.: 1967, *Solar Phys.* **1**, 157.
Rayrole, J.: 1967, *Ann. Astrophys.* **30**, 257.

4. QUIET PHOTOSPHERE AND CHROMOSPHERE

J. M. BECKERS

4.1. Quiet Sun

The 'quiet Sun' is not quiet in the sense that it is free of all disturbance; over the whole surface the fine structures (granulation, spicules) are continuously changing, waves and oscillations occur and large-scale patterns (supergranulation, chromospheric network) evolve. The characteristics of the quiet Sun are that the overall pattern of its fine structures is statistically uniform and/or stationary, that there is no (conspicuous) enhancement or reduction of radiation and that there is no extended magnetic flux concentration (see Figure 3.3). Only at very low resolution ($\gtrsim 10''$) does the quiet Sun really appear quiet.

Gibson, E. G.: 1973, *The Quiet Sun*, NASA SP–303, Washington, D.C.

4.2. Atmospheric Models

A foremost aim of solar physics is to establish a description (model) of the solar atmosphere in terms of the spatial and temporal variation of physical parameters such as electron temperature and density. This aim has not yet been fully realized.

It is usual to assume that a homogeneous (mean) model of the photosphere and low chromosphere can adequately reproduce spatially and temporally unresolved observations, despite the presence of fine structure (granules, network, etc.). The envelopes defined by several recent models of the height variation of the electron temperature and electron number density up to 1000 km are shown in Figure 4.1. These models were derived from observations of many lines and continua, and extended to greater depths by a mixing-length theory of convection.

The layers in which the temperature decreases outwards are termed the photosphere, those in which the temperature increases up to some 10^4 K, the chromosphere. The intermediate region is known as the *temperature minimum*.

Above 1000 km the atmosphere is so heterogeneous that no single-component model can explain the unresolved observations. Two two-component models, invoking spicular and interspicular regions, are shown in Figure 4.1. In model A it is assumed that the cooler component resides in the spicules; in model B the cooler component lies between spicules.

There are indications that a three-component model – including spicular and interspicular regions in the network and an internetwork region – is necessary to explain observations above 2000 km.

Beckers, J. M 1968, *Solar Phys.* **3**, 367.
Lantos, P. and Kundu, M. R.: 1972, *Astron. Astrophys.* **21**, 119.
Vernazza, J. E., Avrett, E. H., and Loeser, R.: 1976, *Astrophys. J. Suppl.* **30**, 1.

Bruzek and Durrant (eds.), Illustrated Glossary for Solar and Solar-Terrestrial Physics. 21–34.

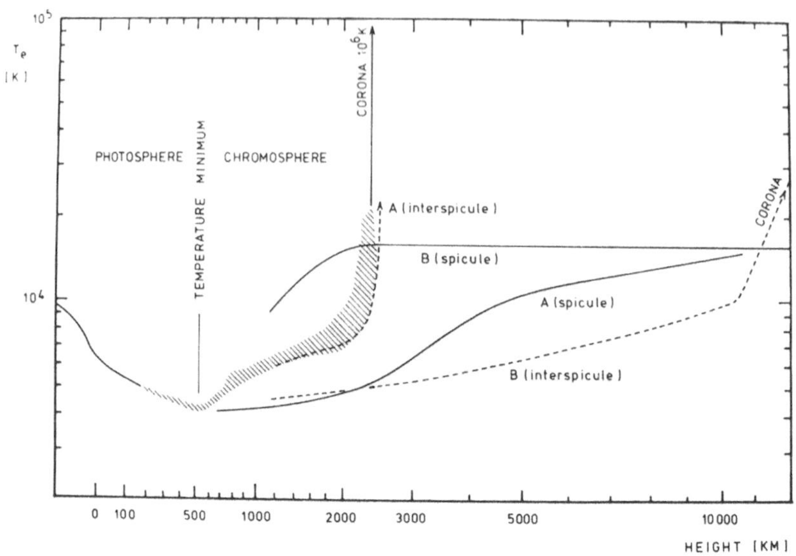

Fig. 4.1a. Run of electron temperature versus height above the base of the photosphere in various solar atmospheric models. The shaded area defines the region within which recent mean photospheric/chromospheric models lie. The curves labelled A and B show two multi-component models.

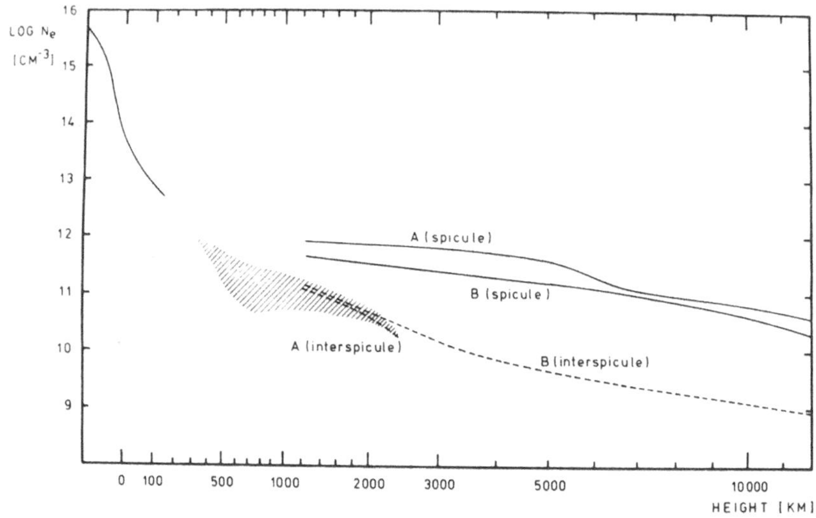

Fig. 4.1b. Run of electron density versus height in various solar models. Symbols as in Figure 4.1a.

4.3. Oscillations

Oscillations of the solar surface are best seen in Doppler shift observations at the centre of the disk but can also be detected in intensity measurements. The average period of the oscillations is 300 s in the photosphere decreasing somewhat with height. The horizontal scale is not as well defined but is ≈10 000 km. The velocity amplitude increases from ≈0.15 km s^{-1} in the low photosphere to ≈0.5 km s^{-1} in the low chromosphere. The vertical phase velocity is ≈100 km s^{-1} and the intensity oscillations lead the velocity by a quarter period.

It has been suggested that the oscillations are evanescent waves excited in the low-temperature solar boundary regions by standing acoustic oscillations which are either high-order non-radial pulsations of the Sun as a whole, or waves trapped in the hydrogen convection zone. Presumably the origin of the oscillations lies in an overstability. This interpretation is favoured by the recent resolution of the photospheric oscillations into discrete modes (Figure 12.2).

In the chromosphere the oscillations can transform back into travelling waves, the energy having tunnelled through the temperature minimum region.

Recently, longer period (up to 2^h40^m) pulsations of the Sun, with mean velocity amplitude of 2 m s^{-1}, have been reported and tentatively identified with low-order acoustic and gravity pulsational modes.

Beckers, J. M. and Canfield, R. C.: 1976, in R. Cayrel and M. Steinberg (eds.), 'Physique des Mouvements dans les Atmosphères Stellaires', CNRS, Paris, p. 207.
Christensen-Dalsgaard, J. and Gough, D. O.: 1976, *Nature* **259**, 89.
Hill, H. A., Stebbins, R. T., and Brown, T. M.: 1975, Proc. 5th Int. Con. Atomic Masses and Fundamental Constants, Paris.
Michalitsanos, A. G.: 1973, *Earth Extraterrest. Sci.* **2**, 125.
Stein, R. F. and Leibacher, J.: 1974, *Ann. Rev. Astron. Astrophys.* **12**, 407.

4.4. Short Period Oscillations

The presence in the solar atmosphere of acoustic or magneto-acoustic waves with periods less than 100 s has often been invoked in order to carry mechanical, non-thermal, energy out of the convection zone. They have been required to deposit this energy progressively in the photosphere by radiative damping, in the chromosphere by weak shocks, and in the corona by viscous and joule losses. The shorter the period the lower is the height at which the energy is dissipated. Ulmschneider's model predicts that 30 s waves give rise to the temperature minimum structure.

Direct observational evidence for such waves is weak since the effects of image motion with time-scales of a few seconds are difficult to disentangle. One interpretation of recent observations of velocity power spectra suggests that waves of all frequencies are present with a total power of some 1–2 km s^{-1} increasing with height. This is consistent with estimates of microturbulent velocities derived from the detailed fitting of line profiles.

Deubner, F. L.: 1976, *Astron. Astrophys.* **51**, 189.
Howard, R. and Livingston, W. C.: 1968, *Solar Phys.* **3**, 434.
Ulmschneider, P.: 1974, *Solar Phys.* **39**, 327.

4.5. Limb Redshift

Compared to their disk-centre wavelengths most weak solar spectral lines appear shifted towards the red end of the spectrum in a region strongly concentrated near the limb. The magnitude of the effect varies from line to line but averages several tenths of a km s^{-1} at the limb. Absolute wavelength measurements indicate that in fact it is due largely to a blueshift at the centre of the disk rather than a redshift at the limb. For strong lines the limb shift is absent.

Many models have been proposed. One suggestion is that the brighter upward moving elements of the photosphere contribute more to the integrated profile at the disk centre than do the darker downward moving elements. Near the limb the projection effects reduce the magnitude of this effect. Another model suggests that the blue-ward pressure shifts of spectral lines are greater near the disk centre than near the limb where the lines are formed higher and at reduced pressure. Both effects are probably contributory. The limb redshift is absent in sunspot umbrae.

Adam, M. G.: 1976, *Monthly Notices Roy. Astron. Soc.* **177**, 687.
Beckers, J. M.: 1977, *Astrophys. J.*, **213**, 900.
Hart, M. H.: 1974, *Astrophys. J.* **187**, 393.

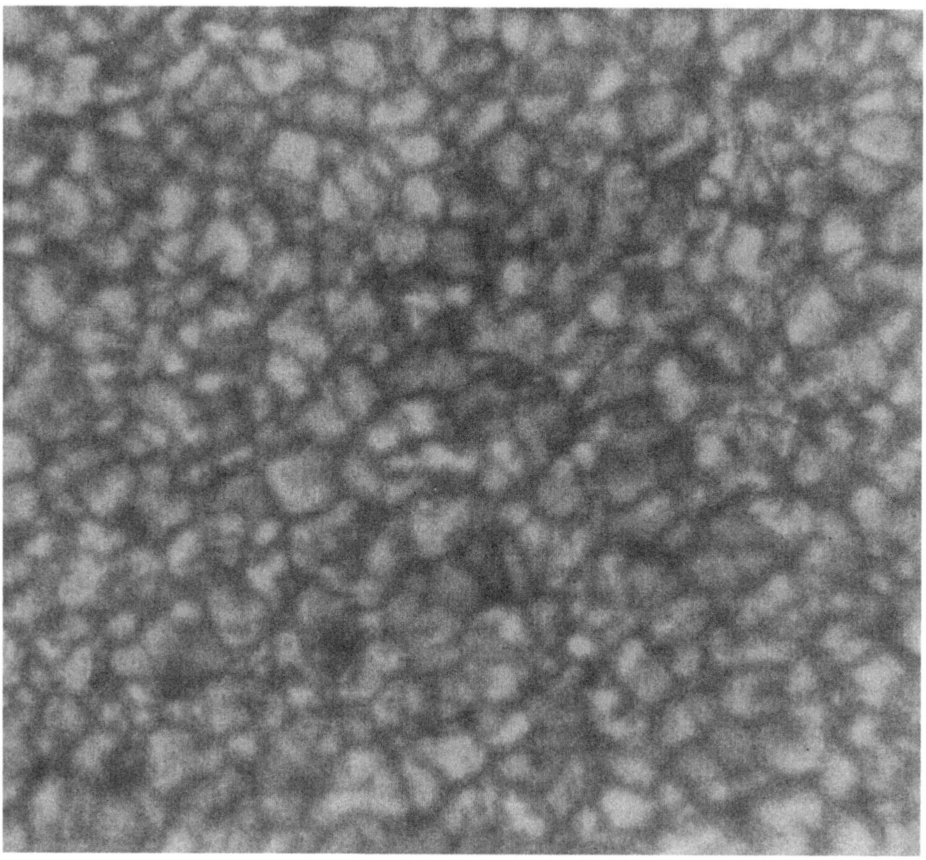

Fig. 4.2. Photospheric granulation near the centre of the quiet Sun with filigree structures in intergranular lanes. Photographed with the vacuum solar telescope at Sacramento Peak Observatory with a 60 Å filter centred at λ6000 Å. (Courtesy S. Koutchmy, September 24, 1975.)

4.6. Granulation

The photospheric granulation consists of a cellular pattern of bright *granules* embedded in a dark *intergranular space*. The majority of granules have diameters 1–2″ and have irregular, frequently polygonal shapes. They appear separated by narrow, dark *intergranular lanes* 0.4″ wide; however, in places, there are relatively large dark areas suggesting the (temporary) absence of one or more granules. These regions should not be confused with pores (which are darker and longer-lived) (Figure 4.2).

Spectra of the granulation show *wiggly lines* (Figure 4.3) which are due to local Doppler shifts of the individual granules: they reveal a vertical upflow in the centre of the granules of 0.4 km s^{-1} and a horizontal outflow of 0.25 km s^{-1}. The granular velocity fluctuations decrease with increasing height in the photosphere and, when extrapolated downward to the height of formation of the continuum, the rms velocity fluctuation is 0.8 ± 0.2 km s^{-1}. Granulation, may be considered to consist of convective elements overshooting the limits of the subphotospheric convection zone.

The lifetime of granules as determined by statistical methods is ≈8 min although individual values up to 15 min are observed. The contrast (brightness) of granules varies considerably during evolution, with height in the atmosphere and between different

Fig. 4.3. Section of solar spectrum (λ6497 Å) with wiggly solar lines and a line of terrestrial origin without wiggles. (Fraunhofer Institut, courtesy W. Mattig.)

granules; rms values of intensity fluctuations of 0.09 to 0.13, referred to the average continuum intensity, have been determined.

Exploding granules are unusually bright and expand at 1.5–2.0 km s^{-1} forming a ring which is eventually fragmented. The whole process takes about 10 min and is a rather common phenomenon.

Abnormal granulation is associated with the filigree; there is evidence that filigree partly overlying the intergranular lanes reduces the contrast of the granulation, thus making it appear abnormal.

Bray, R. J. and Loughhead, R. E.: 1967, *Granulation*, Chapman and Hall, London.
Deubner, F. L. and Mattig, W.: 1975, *Astron. Astrophys.* **45**, 167.
Dunn, R. B. and Zirker, J. B.: 1973, *Solar Phys.* **33**, 281.
Musman, S.: 1972, *Solar Phys.* **26**, 290.

4.7. Photospheric Network

A bright photospheric network appears in spectroheliograms obtained in certain Fraunhofer lines; it is exactly cospatial with the network of non-spot photospheric magnetic fields and coincides in its gross outline with the chromospheric (Ca II) network; it consists, however, of much finer elements which in spectrograms appear as gaps in the absorption lines.

This weakening of the absorption lines is partly due to temperature enhancement and partly to Zeeman splitting. There is evidence that field strengths of ≈2000 G are present in the network in quiet regions, the characteristic size of the magnetic elements being in the range 100–300 km.

Near the solar limb the network becomes visible also in continuum light as photospheric faculae. A *white-light network* was made visible in the centre of the solar disk recently by a time-averaging technique which eliminates the 'noise' produced by the short-lived granulation.

Chapman, G. A. and Sheeley, N. R.: 1968, *Solar Phys.* **5**, 442.
Liu, S.-Y.: 1974, *Solar Phys.* **39**, 297.
Stenflo, J. O.: 1973, *Solar Phys.* **32**, 41.

4.8. Chromospheric Network

The chromospheric network is a large-scale brightness pattern visible in spectroheliograms taken in chromospheric and transition region spectral lines. It appears as a bright structure in the Ca II H, K and infrared lines, Hα line core, and in all lines observed in the vacuum ultraviolet (Figures 4.4 and 7.1). The hydrogen Balmer lines (except Hα core) and the 10830 Å helium line show the network as a dark structure. This network is located at the borders of the photospheric supergranulation and coincides with the regions of local magnetic enhancement (Figures 3.1 and 3.3). The pattern is rather irregular but shows up as a distinct peak in spatial autocorrelation analysis at a wavelength of 33 000 km. The lifetime as determined by autocorrelation techniques is 17 hr.

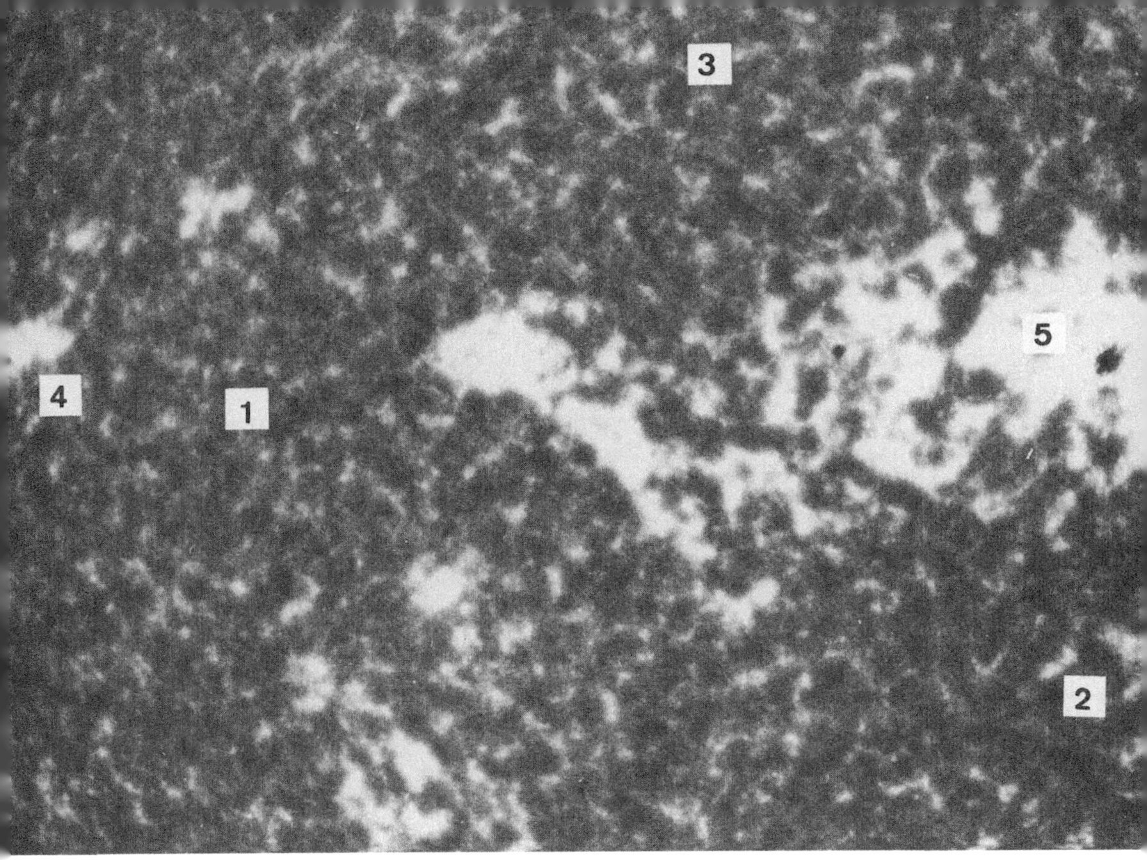

Fig. 4.4. Chromospheric network as seen in the Ca II K line. 1 = network in quiet region, 2 = enhanced network, 3 = old, scattered plages, 4 = young AR (emerging flux region) with bright plage, 5 = mature plage region. (Fraunhofer Institut Anacapri Observatory.)

Hα network cells are outlined by chains and rosettes at the vertices of several cells. Their location corresponds to that of the bright coarse structures (≈5000 km) seen in the H and K Ca II lines, flocculi, and to that of photospheric magnetic field enhancement. In the quietest regions of the solar surface the network is barely distinguishable, the rosette structures being absent or very incomplete. There is a smooth transition to well-defined network outlined by rosettes, to *enhanced network* in which the rosettes are grouped more densely and contain brighter mottles, *plagettes*, and from them to active region plages (Figure 4.4). In Hα the increasing activity is accompanied by an increasing length of the dark mottles, merging with fibrils, which cover more and more of the network cell interiors.

Bray, R. J. and Loughhead, R. E.: 1974, *The Solar Chromosphere*, Chapman and Hall, London.
Simon, G. W. and Leighton, R. B.: 1964, *Astrophys. J.* **140**, 1120.

4.9. Supergranulation

The supergranulation is a system of large-scale (32 000 km) velocity cells seen best in the photosphere away from the centre of the solar disk as a pattern of horizontal motions.

This pattern does not vary significantly over the quiet solar surface nor with phase of the solar cycle. Near active regions the cell size seems to increase by about 10%.

The horizontal motions are typically 0.3–0.4 km s^{-1}, and downward vertical motions ≈0.1–0.2 km s^{-1} occur principally in magnetic regions which border the supergranulation cells. Upward motions at the centre of the cells seem to be even weaker. The velocities decrease with height. The lifetime of the velocity cells is uncertain, perhaps 1–2 days.

The cells are presumably convective in origin reflecting a large preferred cell size in the convective zone. Since the cells are virtually invisible in integrated light there is little or no temperature difference across them.

There is a close connection between the chromospheric network and the supergranulation in the sense that the former lies at the boundaries of the supergranule cells and probably results from the accumulation of magnetic flux at the borders of the horizontal flow.

Simon, G. W. and Leighton, R. B.: 1964, *Astrophys. J.* **140**, 1120.
Beckers, J. M. and Canfield, R. C.: 1976, in R. Cayrel and M. Steinberg (eds.), 'Physique des Mouvements dans les Atmosphères Stellaires', CNRS, Paris, p. 207.

4.10. Mottle

Mottle is a general expression used to describe details in monochromatic images of the chromospheric network. At high spatial resolution the components of the network appear usually as somewhat elongated features, *fine mottles*, with dimensions of about 1 × 10″. Their lifetime is ⩽10 min. Almost equal numbers show upward and downward velocities. Fine mottles are best seen in the wing of Hα. At Hα line centre they appear bright (bright mottles) if they are low-lying (700–3000 km above the base of the photosphere) and dark (dark mottles) if higher (3000–10 000 km) in the atmosphere. Generally only the low-lying mottles are seen in Ca H K line which always appear bright. For mottles in Hα see Figures 7.6 and 7.7.

High-reaching mottles are seen beyond the limb as spicules. The physical conditions in mottles are difficult to determine directly but are probably similar to those of spicules. The discrepancy in inferred velocity has been attributed to a seeing effect.

Fine mottles are commonly grouped into clusters, rosettes or *flocculi*, which appear as entities, *coarse mottles*, when observed at low spatial resolution. (See Figure 4.5.)

Bray, R. J. and Loughhead, R. E.: 1974, *The Solar Chromosphere*, Chapman and Hall, London.

4.11. Rosette and Chain

These are common arrangements of fine mottles or spicules, best seen in Hα wing filtergrams taken at the centre of the solar disk, in which a number of dark elongated features radiate outwards from a common centre or are aligned in a double row (chain). At Hα line centre, bright mottles appear at the centre of a rosette and between the dark mottles. Typically, the overall size is some 10 000 km, and some 40 mottles make up the rosette. However, many rosettes are broken or only partially completed (Figure 7.7).

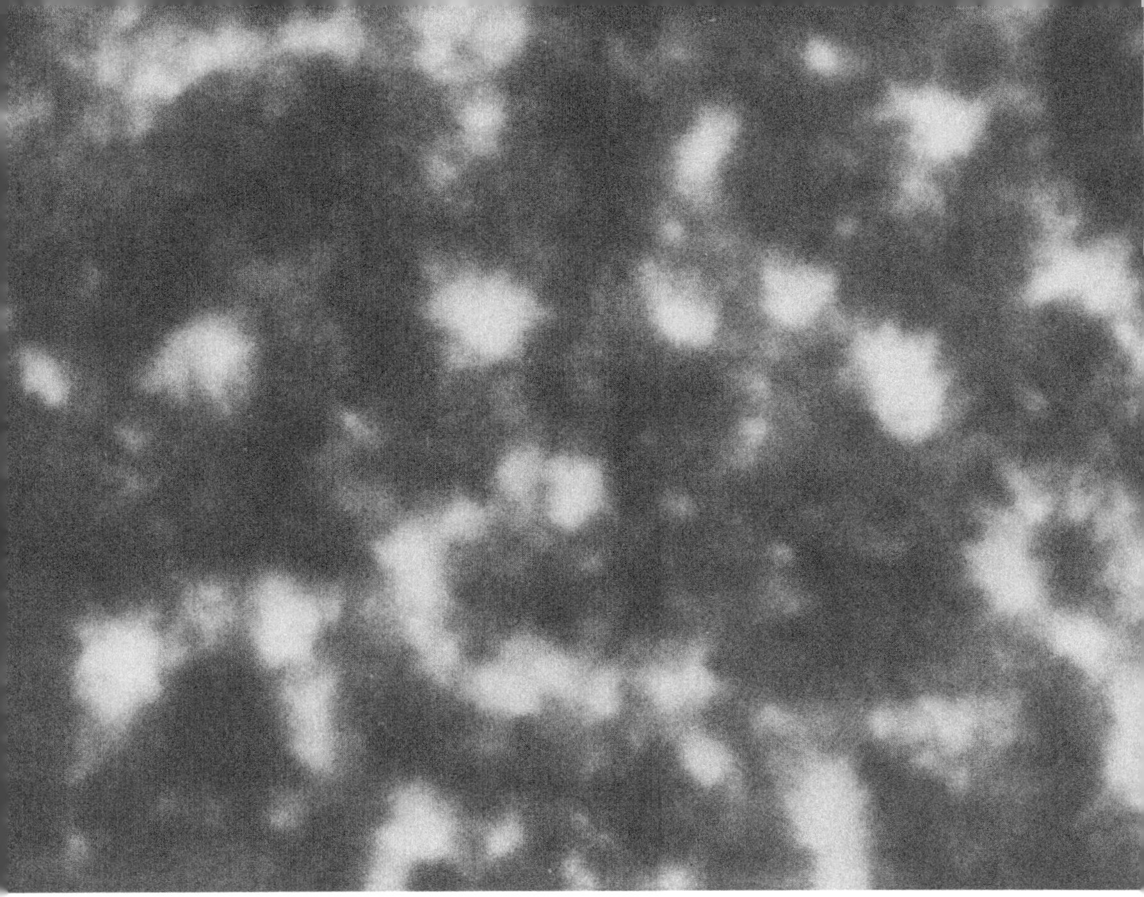

Fig. 4.5. Ca II K Flocculi. (Sacramento Peak Observatory.)

When seen near the solar limb the rosette mottles all point out towards the limb forming a *bush*. Right at the limb the corresponding spicule configuration has been called a *porcupine structure*. (See Figures 4.6 and 7.7.)

Beckers, J. M.: 1972, *Ann. Rev. Astron. Astrophys.* **10**, 73.
Bray, R. J. and Loughhead, R. E.: 1974, *The Solar Chromosphere*, Chapman and Hall, London.

4.12. Chromospheric Grain

There has been no systematic description of the chromosphere inside supergranulation cells so it is possible that the features mentioned under this heading may not all refer to the same phenomenon.

In the CA H and K lines, cell interiors are filled with oscillating elements. They are <1000 km in size and have a quasi-period of 200 s. The brightening is particularly prominent in the violet wing (H_{2v}, K_{2v}) causing noticeable *bright cell points*, grains or *inner cell dots* in spectroheliograms lasting 30–100 s. These are accompanied by progressive intensity waves in the upper photosphere, as revealed in the wings of Ca II K, which gave rise to dark and bright *whiskers* in the spectrum.

Fig. 4.6. Spicule bushes as seen in the wing of the Hα line near the solar limb. Also note the network outlined by the bushes. (Sacramento Peak Observatory.)

In Hα, supergranulation cell interiors when not obscured by fibrils show a brightness and velocity pattern with a scale of ≈1000 km that has been described as a *chromospheric granulation*. The upward velocity shifts show clearly at Hα–0.5 Å, the displaced features also being known as grains.

The mechanism is uncertain but may be simply the penetration of the photospheric velocity field through the temperature minimum via waves or shocks.

Beckers, J. M. and Artzner, G.: 1974, *Solar Phys.* **37**, 309.
Giovanelli, R. G.: 1974, *Solar Phys.* **37**, 301.
Liu, S.-Y.: 1974, *Astrophys. J.* **189**, 359.

4.13. Spicule

Spicules are rapidly changing, predominantly vertical, spike-like structures in the solar chromosphere observed beyond the limb. They are best visible in the H Balmer lines but are also observed in other chromospheric lines and perhaps in the continuum. They have diameters ≈1000 km, lengths ≈6000–10 000 km, temperatures $\approx 1-2 \times 10^4$ K, and electron densities $\approx 3 \times 10^{10} - 3 \times 10^{11}$ cm^{-3}. About 10^6 are seen on the Sun at any one time. They appear to be 'ejected' from the low chromosphere reaching a speed ≈20–30 km s^{-1} and a height ≈9000 km and then fall back and/or fade along their full

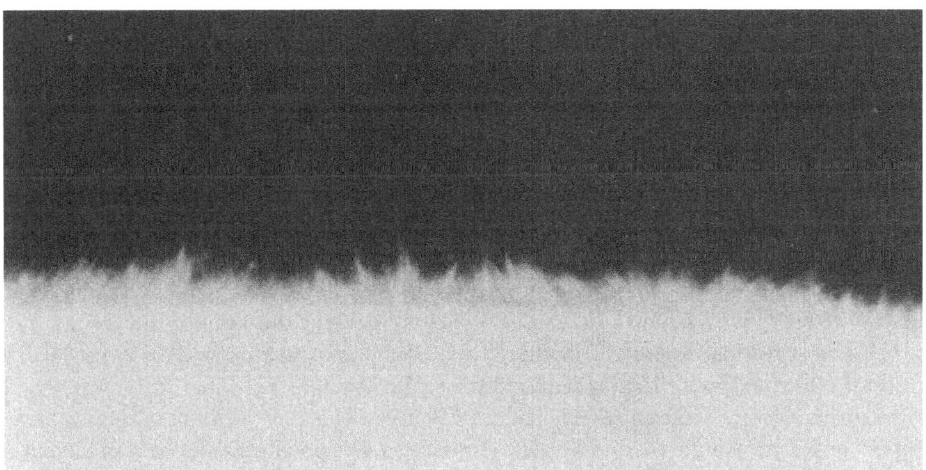

Fig. 4.7. Spicules at the solar limb observed in the centre of the Hα line. (Sacramento Peak Observatory.)

length. The total lifetime is 5–10 min. They apparently have substantial rotational velocities. See Figure 4.7.

Spicules are distributed non-uniformly across the solar disk, most, if not all, being located at the supergranule boundaries. They are identified with high-reaching mottles seen against the solar disk. See Figure 4.6.

The group behaviour of spicules as seen outside the solar limb in which all spicules over an ≈140 000 km arc along the limb appear to have the same orientation has been described as a *wheat field pattern.*

The origin of spicules is probably to be found in the interaction of the photospheric granules with the magnetic field at the supergranule boundaries, with a possible contribution by strong conductive energy flux from the corona at these locations.

Beckers, J. M.: 1972, *Ann. Rev. Astron. Astrophys.* **10**, 73.
Bray, R. J. and Loughhead, R. E.: 1974, *The Solar Chromosphere*, Chapman and Hall, London.

4.14. Interspicular Region

The region between solar spicules has ill-defined properties; probably there is no unique region of such a type. One might expect for example, the interspicular region at the network boundary to be different from that in the interior of the network cells. It is generally believed that the interspicular temperatures at a specific height are much higher than that of spicules at that same height. Coronal temperatures (10^6 K) have been suggested.

Beckers, J. M.: 1972, *Ann. Rev. Astron. Astrophys.* **10**, 73.

4.15. Macrospicule

Macrospicules are jets of chromospheric material resembling small surges or giant spicules. They are limited to the region of (polar) coronal holes and are visible in H Lα, He II 304 Å as well as in transition region lines (C II, C III, O IV) up to $T \approx 2 \times 10^5$ K.

They appear to rise up to a length of 5″–50″ and then either to fall back along their own axis or simply to fade from view. Velocities, both rising and falling are 10–150 km s^{-1}; observed lifetimes range between 8 and 45 min; the diameters are 5″–15″. The electron density of macrospicules is $\approx 10^{10}$ cm^{-3} (compared to 1.3×10^{11} cm^{-3} of a normal Hα spicule) and the optical thickness in Hα is only $\frac{1}{20}$ of the value in a normal spicule. The energy required for a macrospicule is $\approx 3 \times 10^{26}$ erg, i.e. about 100 times more than is required for an Hα spicule.

Macrospicules are increasingly inclined away from the pole as a function of increasing position angle measured from the pole (similar to Hα polar spicules and polar plumes) indicating that their trajectories are defined by the – presumably unipolar – magnetic field in the polar cap.

Bohlin, J. D. *et al*: 1975, *Astrophys. J.* **197**, L133.
Withbroe, G. L. *et al*,: 1976, *Astrophys. J.* **203**, 528

4.16. Chromospheric Bubbles or 'Bulles'

These are roundish features observed in Hα moving upwards from the limb chromosphere with velocities up to $\gtrsim$100 km s^{-1}. After an average lifetime of 1.8 min they fade having reached heights in the range 5300–14 000 km. Their average diameter is 1800 km. The number of bubbles existing at any time between 5000 and 10 000 km is estimated to be $\approx$430. An association with either spicules or with chromospheric grains is suggested by Mouradian.

Mouradian, Z. and Simon, G.: 1975, *Solar Phys.* **42**, 311.

4.17. Emission Shell

This is the emission region seen about 1500 km above the solar limb which is clearly detached from the solar limb. It is seen in the helium D_3 (White) and in the wing of the Hα line (Bray and Loughhead). The darkening between this emission shell and the solar limb is referred to as the *dark band*.

Bray, R. J. and Loughhead, R. E.: 1974, *The Solar Chromosphere*, Chapman and Hall, London.
White, O. R.: 1963, *Astrophys. J.* **138**, 1316.

4.18. Flash Spectrum

The flash spectrum is the emission spectrum of the solar chromosphere observed at a total solar eclipse often with a slitless spectrograph. It reveals a wealth of chromospheric emission lines. The motion of the lunar limb across the chromosphere provides a means of determining the height variation of the chromospheric emission with better spatial resolution than is possible with the best resolved images.

Van de Hulst, H. C.: 1953, in G. P. Kuiper (ed.), *The Sun*, University of Chicago Press, Chicago, p. 216.

4.19. Spectroheliograph

In general a spectroheliograph is an instrument capable of giving an image of the Sun in a wavelength band of <1 Å. In particular, it is a spectrograph equipped with an exit slit in addition to the entrance slit. The exit slit is located at the wavelength at which the image is desired. The image is obtained by moving the spectrograph, including both slits, with respect to the solar image and the photographic plate. In this way a monochromatic image is recorded called a *spectroheliogram*.

In previous years, spectroheliographs have been extensively used in the core or wings of strong chromospheric lines (e.g. Balmer lines, Ca II H and K) to take photographs of the chromosphere at different heights and of chromospheric phenomena such as filaments and flares. For these purposes the spectroheliographs are now largely replaced by birefringent filters. They are still used – with two exit slits – for the production of Leighton magnetograms and dopplergrams. The photographic film may be replaced by photoelectric devices using for instance a diode array. Special types of spectroheliograms have been used also in satellite equipment for EUV line photographs of the Sun.

Title, A.: 1966, *Selected Spectroheliograms*, California Inst. Technology, Pasadena.

4.20. Birefringent Filter

This is a colour filter using the rapid variation with wavelength of the retardation of uniaxial crystals like quartz and calcite to obtain a very narrow spectral transmission profile. Two types of filters are in use – the Lyot filter (also called the Lyot-Öhman filter) and the Šolc filter. Images obtained with narrow band filters are called *filtergrams*.

The *Lyot filter* consists of a series of 'elements' each of which (in its simplest configuration) consists of a birefringent crystal placed between two polarizers. Such an element gives a transmission profile vs. wavelength which is sinusoidal (≡ *channel spectrum*). A series of such elements with factors of 2 difference between their thicknesses and with proper phasing of their channel spectra gives the Lyot filter. Filter bandwidths of ≈0.1 Å have been realized. Modifications of the basic concept include wide angular acceptance angle configurations and tunability. So-called *Universal Birefringent Filters* are tunable over a wide wavelength region (e.g. 4000–7000 Å).

The *Šolc filter* consists of a stack of equal thickness birefringent plates placed between two polarizers. These filters have a smaller angular field than modern Lyot filters and are not so easily tunable. Their main advantage lies in their higher transmission resulting from the use of only two polarizers.

Evans, J. W.: 1949, *J. Opt. Soc. Am.* **39**, 229.
Evans, J. W.: 1958, *J. Opt. Soc. Am.* **48**, 142.
Lyot, B.: 1944, *Ann. Astrophys.* **7**, 31.
Šolc, I.: 1965, *J. Opt. Soc. Am.* **55**, 621.

5. TRANSITION REGION

C. JORDAN

5.1. Transition Region

This term is generally used to indicate the region where the temperature rises from chromospheric values of $\lesssim 10^4$ K to values close to those in the inner corona. This region is observable mainly through the EUV emission lines.

Because thermal conduction plays the dominant role in determining the temperature gradient above $T_e \approx 10^5$ K (at least in quiet Sun areas), the term is often used synonymously with a region of constant conductive flux. However, it is alternatively used to indicate a region of steep temperature gradient and then it would include the region between 10^4 K and 10^5 K. For example, typical models based on spatially averaged intensity observations give a thickness of only 30 km between $T_e \approx 3 \times 10^4$ K and 3×10^5 K, compared with a thickness of 2500 km between $T_e \approx 2 \times 10^5$ K and 10^6 K.

Observations are consistent with the transition region existing as a sheath around inhomogeneities, such as spicules, rather than as a horizontal layer. The average height at $T_e \approx 10^5$ K in the quiet regions has been measured as 1700 ± 800 km.

Observations from OSO-6 have shown that the temperature gradient in material at temperatures $>10^5$ K in coronal holes is an order of magnitude smaller than that in quiet-Sun areas. From ATM, the height difference between the Ne VII line ($T_e \approx 5.5 \times 10^5$ K) and the Lyman continuum was measured directly, giving 8000 km in a polar hole as against $\approx$1700 km in quiet-Sun models.

In active regions the temperature gradient at these temperatures is a factor of five steeper than in quiet regions whereas directly over sunspots, at $T_e \approx 5 \times 10^4$ K, it is a factor of ten lower than in plages and the density is also lower.

Models of the transition region in quiet-Sun regions have been made by several methods. Above $T_e \approx 10^5$ K the assumptions of constant conductive flux with no radiation losses and deposition of energy only at the top of the atmosphere have been used. Alternatively, models can be made for $T_e \gtrsim 2 \times 10^4$ K by using the observed emission measure distribution as a function of temperature, together with the equations of conductive flux and hydrostatic equilibrium and a pressure boundary condition. Although little energy deposition and radiation losses appear to occur at $T_e \gtrsim 10^5$ K, below this temperature the conducted energy and any energy deposited must be balanced, and radiation losses become important. If the restriction of hydrostatic equilibrium is removed then models in which the transition region is thicker can result. Recent work has been concerned with improving earlier models by taking into account the spatial variation of the EUV emission, and the rôle of the magnetic field.

Brueckner, G. E. and Nicolas, K. R.: 1973, *Solar Phys.* **29**, 301.
Burton, W. M., *et al.*: 1973, *Astron. Astrophys.* **27**, 101.
Dupree, A. K.: 1972, *Astrophys. J.* **178**, 527.
Gabriel, A. H.: 1976, *Phil. Trans. Roy. Soc. London* **A281**, 339.
Huber, M. C. E., *et al*: 1974, *Astrophys. J.* **194**, L115.
Kopp, R. A. and Kuperus, M.: 1968, *Solar Phys.* **4**, 212.

Bruzek and Durrant (eds.), Illustrated Glossary for Solar and Solar-Terrestrial Physics. 35–38.

Loulergue, M. and Nussbaumer, H.: 1974, *Astron. Astrophys.* **34**, 225.
Munro, R. H. and Withbroe, G. L.: 1972, *Astrophys. J.* **176**, 511.
Noyes, R. W., Withbroe, G. L., and Kirshner, R. P.: 1970, *Solar Phys.* **11**, 388.
Shmeleva, O. P. and Syrovatskii, S. I.: 1973, *Solar Phys.* **33**, 341.

5.2. Di-Electronic Recombination

Di-electronic recombination occurs in two stages. In the first stage an incident electron excites an ion to an energy corresponding to that of an auto-ionising level in the recombined ion, into which it can be captured. In the second stage, for recombination to be effective, this excited level must undergo spontaneous radiative decay through a transition of an inner electron to a lower level which is not liable to rapid ionisation. The reactions can be written as

$$X^{+z}(i) + e(E, l') \rightleftharpoons X^{+(z-1)}(j, nl)$$
$$X^{+(z-1)}(j, nl) \longrightarrow X^{+(z-1)}(k, nl) + h\nu,$$

where $X^{+z}(i)$ is a z-times ionised atom in the state i (usually the ground configuration), $X^{+(z-1)}(j, nl)$ is a $(z-1)$-times ionised atom in a (usually) doubly excited state (j, nl).

If auto-ionisation takes place before the second stage then recombination has not occurred. However, because of detailed balance large auto-ionisation rates imply large di-electronic capture rates and a high level population builds up. Thus levels which have an auto-ionisation rate larger than the spontaneous radiative decay rate contribute most to the total di-electronic recombination.

For most coronal ions (but not for H I-like or He I-like ions) di-electronic recombination is two orders of magnitude larger than radiative recombination. Its inclusion in the coronal ionisation equilibrium equation raised the temperature at which a given coronal ion was predicted to be most effectively produced and removed a long-standing discrepancy between the coronal temperatures derived from line widths and from the relative intensities of lines from adjacent stages of ionisation. More recent work has considered the density dependence of the process, which arises in the second stage of the recombination.

Burgess, A.: 1964, *Astrophys. J.* **139**, 776.
Summers, H. P.: 1974, *Monthly Notices Roy. Astron. Soc.* **169**, 663.

5.3. Intersystem or Intercombination Lines

Intersystem or intercombination lines are those from transitions between states of different spin, made possible through the breakdown of LS coupling. This breakdown increases with high atomic number but intersystem lines from a wide variety of elements and ions are observed in the solar spectrum, e.g.

$$\text{O I } 2p^4\ ^3P_1 - 2p^3 3s\ ^5S_2,\ \text{O V } 2s^2\ ^1S_0 - 2s2p\ ^3P_1,$$
$$\text{Fe XVII } 2p^6\ ^1S_0 - 2p^5 3s\ ^3P_1,\ \text{Fe XXV } 1s^2\ ^1S_0 - 1s2p\ ^3P_1.$$

Most of the intersystem transitions are weak compared with permitted transitions when spectra are taken at the centre of the solar disk, because their collisional excitation rates are smaller than those for permitted transitions. In low ions the spontaneous radiative transition probabilities for these lines can be small enough for collisional de-excitation processes to compete in de-populating the excited level. Then the relative intensity of permitted and intersystem transitions can be used to measure the local electron density.

Herzberg, G.: 1944, *Atomic Spectra and Atomic Structure*, Dover Publications, New York.
Jordan, C.: 1973, *Nuclear Instruments and Methods* **110**, 373.

5.4. Satellite Lines

A transition between two levels with given quantum numbers in the presence of an electron of much larger principal quantum number will have a wavelength which is very similar to that of the same transition in the absence of the high level electron. When the extra electron is in a low orbit the separation in the wavelength of the transitions is large enough for the line from the system with the extra electron to appear as a 'satellite' to the long wavelength side of the basic transition. For example, the transition $1s^2 - 1s2p$ in the helium-like ions has adjacent to it transitions of the type $1s^2 2s - 1s2s2p$ and $1s^2 2p - 1s2p^2$.

Two types of satellite lines are observed in the solar soft X-ray spectrum. Most satellite lines (adjacent to H I-like and He I-like resonance lines) originate from the process of di-electronic recombination, when the excited state stabilises by spontaneous radiative decay,

$$\text{e.g.}\quad 1s^2 + e \rightleftharpoons 1s2pnl$$

$$1s2pnl \longrightarrow 1s^2 nl + h\nu \text{ (satellite).}$$

However, some satellite lines are observed (adjacent to He I-like resonance lines) from excited levels which cannot be reached by di-electronic capture because no continuum of the same parity and angular momentum exists. These lines are formed by electron impact excitation of an inner electron in the Li I-like ion,

$$\text{i.e.}\quad 1s^2 2p + e \longrightarrow 1s2p^2 + e.$$

Provided the lines are optically thin, the relative intensity of the di-electronic satellite lines to that of the resonance lines depends in a given ion on T_e^{-1} and increases with increasing ion charge. In ionisation equilibrium the di-electronic satellites are usually stronger than the inner shell impact satellites, but the latter can become important where the material is rapidly ionising. The inner shell satellites also become relatively more important in higher ions.

Bhalla, C. P., Gabriel, A. H., and Presnyakov, L. P.: 1975, *Monthly Notices Roy. Astron. Soc.* **172**, 359.
Doschek, G. A.: 1972, *Space Sci. Rev.* **13**, 765.
Gabriel, A. H. and Jordan, C.: 1969, *Nature 221*, 947.

5.5. Grazing-Incidence Optics

In order to make observations below about 300 Å with reflecting optics it becomes necessary to use systems where the angle of incidence is very small (say less than 15° at the longer wavelengths), since the critical angle for total external reflection decreases as the wavelength decreases. Such optical systems are usually referred to as grazing- or glancing-incidence optics. The main types of telescopes used are the *Wolter telescopes*; they consist of either a paraboloid-confocal-hyperboloid system (Wolter Type I) or a paraboloid and a hyperboloid used externally (Wolter Type II). Both full-figure of revolution telescopes and sectors of these have been used for solar observations.

Firth, J. G., *et al.*: 1974, *Monthly Notices Roy. Astron. Soc.* **166**, 543.
Giacconi, R., *et al.*: 1955, *J. Opt. Soc. Am.* **55**, 345.
Wolter, H.: 1952, *Ann. Physik* **10**, 94.

5.6. Crystal Spectrometers

In the soft X-ray region, say between 1.5 Å and 25 Å, spectra are usually obtained using the diffraction of X-rays from a crystal. Spectral resolution is achieved by rotating the crystal, making use of Bragg's law of refraction. The spectrometers are referred to either as crystal spectrometers or *Bragg crystal spectrometers*, and have been used for solar observations since 1965. Proportional counters are usually used as detectors and spatial resolution can be obtained by using mechanical collimators.

Blake, R. L., *et al.*: 1965, *Astrophys. J.* **142**, 1.
Brabban, D. H. and Glencross, W. M.: 1973, *Proc. Roy. Soc. London* **A334**, 231.

6. SOLAR CORONA

S. KOUTCHMY

6.1. White-Light Corona

The white-light corona becomes visible during total solar eclipses. Outside eclipses it can be observed at 2–9 $R_{\odot}$ by balloon- or satellite-borne, externally occulted coronagraphs (OSO-7, Skylab). It consists of two components: the K-corona and the F-corona.

The *K-corona, electron corona* or *continuum corona* is produced by Thomson scattering of photospheric radiation on the free electrons of the highly ionized coronal plasma. Its spectrum is continuous since the Fraunhofer lines of the photospheric spectrum are almost completely smeared out by large Doppler shifts due to the fast

Fig. 6.1. *K-corona* photographed during the June 30, 1973 total eclipse using a radial neutral filter (Institut d'Astrophysique CNRS, Paris). The contrast of the structures has been enhanced by special photographic processing. *H* helmet streamer, *R* condensation, *P* plumes, *S* sharp-edge streamer, *F* fine rays. A Hα photograph of the Sun taken at the same time (Sacramento Peak Observatory) is inserted.

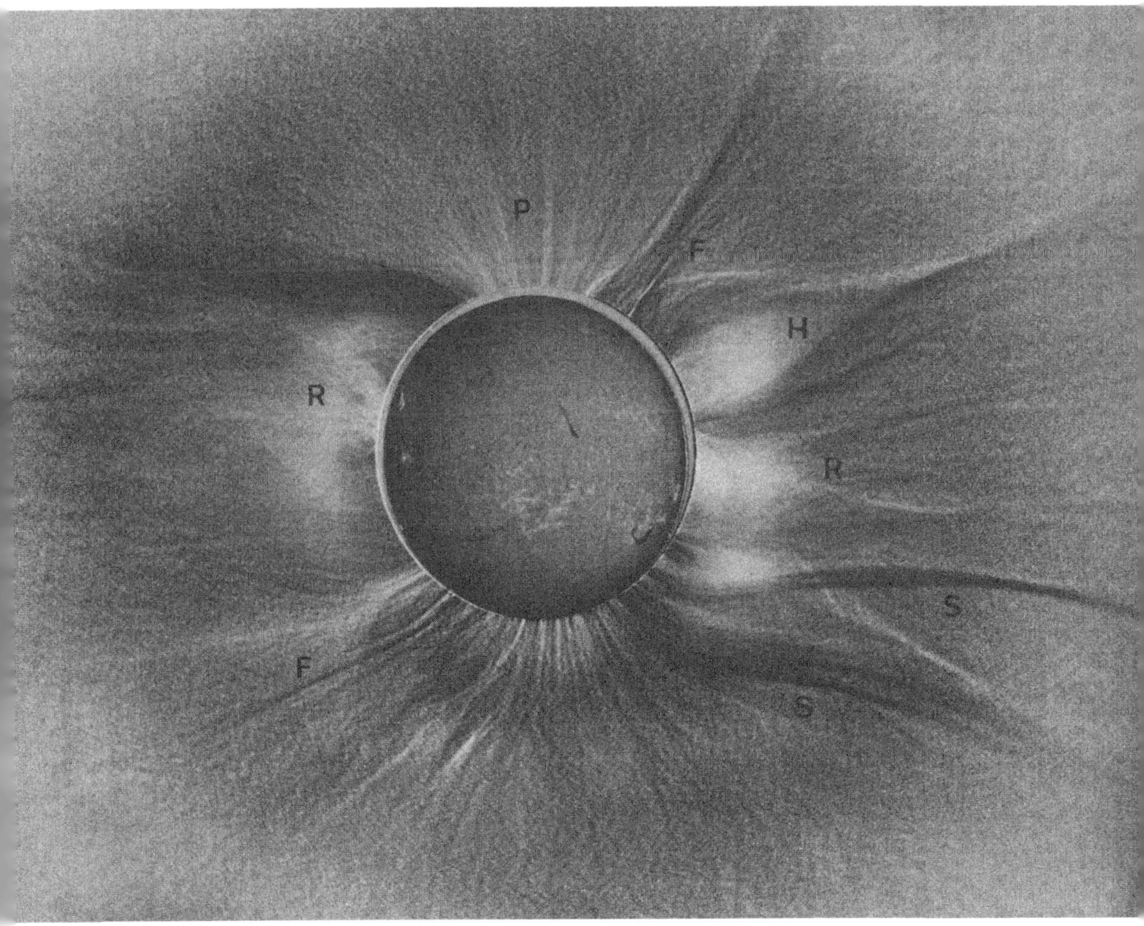

Bruzek and Durrant (eds.), Illustrated Glossary for Solar and Solar-Terrestrial Physics.
39–52.

electrons of the 10^6 degree corona. The scattered radiation is strongly linearly polarized. The K-corona is very inhomogeneous, containing a number of characteristic structures such as streamers, arches, plumes and fine rays (Figure 6.1). Average electron densities are several 10^8 cm^{-3} in the quiet inner corona; they are enhanced 5–20 times in coronal structures compared to the undisturbed surroundings, in coronal condensations up to 10^{10} cm^{-3}.

The *F-corona*, *Fraunhofer corona* or *dust corona* is due to photospheric light scattered on dust particles surrounding the Sun. Since the particles are slow it shows the Fraunhofer lines. The scattered radiation is unpolarized. The extension of the F-corona into interplanetary space appears as *Zodiacal light.*

Blackwell, D. E., Dewhirst, D. W., and Ingham, M. F.: 1967, *Adv. Astron. Astrophys.* 5, 1.
Newkirk, G., Jr.: 1967, *Ann. Rev. Astron. Astrophys.* 5, 213.

6.2. Ellipticity or Flattening

The *ellipticity coefficient* ϵ characterizes the shape of the isophotes of the white-light corona. It is defined as

$$\epsilon = \frac{r_e}{r_p} - 1;$$

r_e, r_p are respectively the equatorial and polar distances of the isophotes from the centre of the solar disk according to van de Hulst's definition. Ludendorff used instead the means of the values at three positions: 0°, ± 22°.5 and 90°, 90° ± 22°.5 respectively. The ellipticity depends on the distance from the Sun and on the phase of the solar cycle. It is small (0.05) during solar maximum (*maximum corona*) and large (≈0.25) during solar minimum (*minimum corona*).

van de Hulst, H. C.: 1953, in G. P. Kuiper (ed.), *The Sun*, Univ. of Chicago Press, Chicago, p. 285.

6.3. Coronal Streamers

Coronal streamers are characteristic, approximately radial structures of 3–10 times enhanced electron density of the K-corona extending beyond 0.5–1 $R_\odot$ to distances up to 10 $R_\odot$ (=solar radii). Various types have been distinguished.

Active region streamers form above young active regions and show marked evolution in a few weeks. They are seen most clearly in the spot zones as the solar cycle advances. The streamers in the equatorial regions at solar minimum (*equatorial streamers*) are also very characteristic. Active region streamers apparently straddle the coronal enhancement above the active region and extend outwards for 3–4 $R_\odot$ as a series of *bushes* or *fans* of *thread-like streamers* or *rays*. The most conspicuous form *large streamers* which are cusp-like at the base (sometimes referred to as 'bulb'), narrow to a 'throat' or 'neck' at 2–3 $R_\odot$ above the base and then extend in a straight, slightly diverging 'stalk' (Figure 6.2). In the stalk coronal material moves outwards by supersonic expansion.

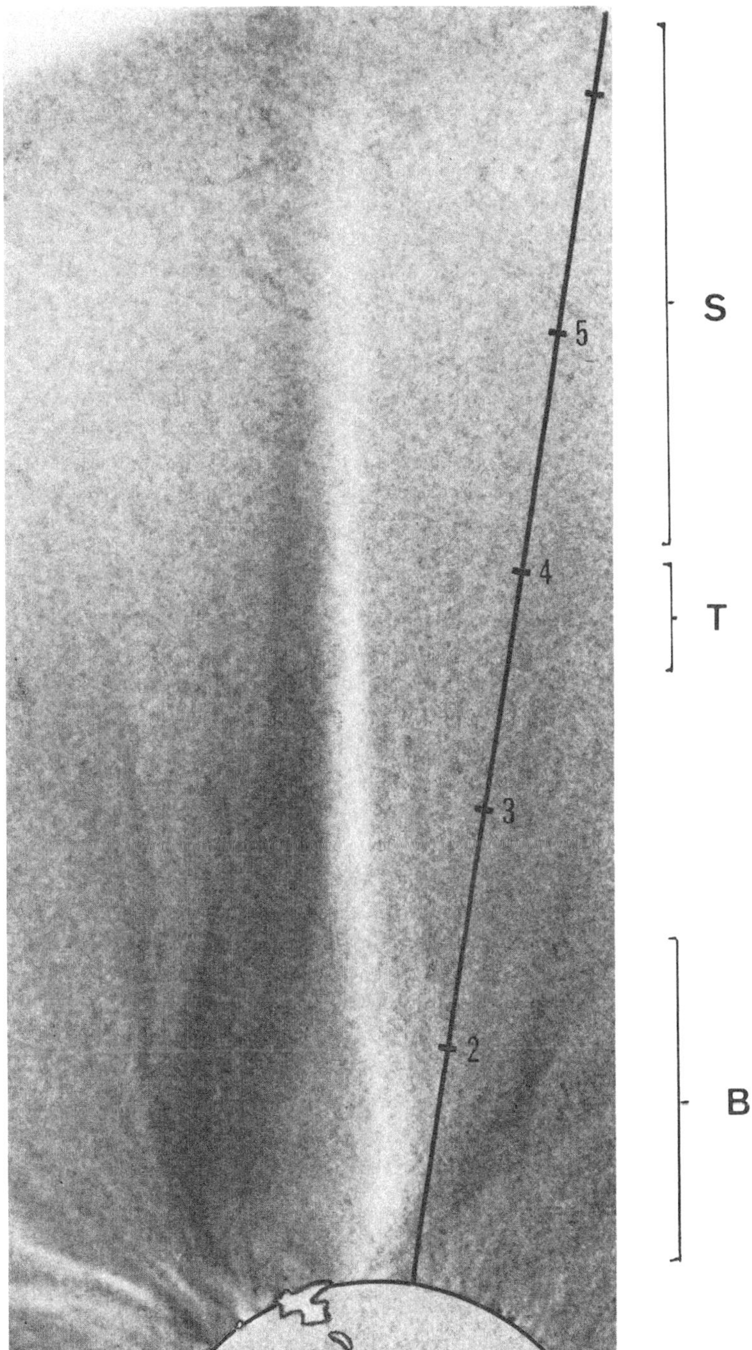

Fig. 6.2. *Large streamer* at the March 7, 1970 eclipse. The contrast of the photograph was enhanced by employing a radial neutral filter and by dodging the print. *S* stalk, *T* throat or neck, *B* bulb or cusp. (S. Koutchmy.)

Helmet streamers lie above quiescent prominences or extended bipolar regions and live for many months. They are best seen over the chains of filaments that form in higher latitudes in mid-solar cycle. They have broad bases ($\leqslant 1\ R_\odot$) which appear as the super-position of a multitude of coronal arches contracting to a pointed top 1–2 $R_\odot$ above the solar limb (Figure 6.1).

The term *narrow ray*, *mini-streamer* and *screw-like structure* presumably all refer to the much narrower, isolated streamers seen outside blade-like systems. Their nature is as yet unknown but some appear tied to plage filaments.

There is growing evidence that most, if not all, streamers have the same basic structure: a rounded base of closed field lines surmounted by a blade of open field lines stretched into almost parallel rays. Seen end-on it appears as a large or helmet streamer, seen from the side it forms a fan. There appears to be a tangential discontinuity in the electron density, which is seen sometimes as a sharp edge (*sharp-edge streamer*) (Figure 6.1).

An evolutionary connexion has also been proposed. Streamers form when the plasma in a coronal condensation becomes hot enough to burst open the closed field loops to form a plane neutral (current) sheet above the photospheric magnetic inversion line. An active region or helmet-type streamer appears depending on the local conditions. A helmet streamer can remain for up to 5 solar rotations.

Active region and helmet streamers are associated with enhanced solar wind density and flux. Streamers are the source of the slowly varying component of the metric radio emission and they are also related, in a manner which is still unclear, to Type III bursts.

Bohlin, J. D.: 1970, *Solar Phys.* **12**, 240; **13**, 153.
Koutchmy, S.: 1972, *Solar Phys.* **24**, 373.
Newkirk, G., Jr.: 1967, *Ann. Rev. Astron. Astrophys.* **5**, 213.
Steinberg, J. L.: 1975, in: Chiuderi *et al.* (eds), *Proc. First European Solar Meeting*, Florence, p. 56.

6.4. Polar Plumes

Polar plumes or *polar rays* are characteristic ray-like structures of the undisturbed K-corona which are well observed in the polar regions of the Sun during solar minimum. They are relatively short-lived, lifetime ≈ 15 hr. Plume-like structures are also seen in equatorial regions presumably in coronal holes. Close to the solar limb fine structures have been observed in plumes (*coronal spikes*). The distribution of plumes is statistically associated with the chromospheric network and surface magnetic fields. Around the poles the plumes apparently outline the polar magnetic field of the Sun (Figures 6.1, 6.3).

Koutchmy, S. and Stellmacher, G.: 1976, *Solar Phys.* **49**, 253.
Newkirk, G., Jr. and Harvey, J.: 1968. *Solar Phys.* **3**, 321.

6.5. Coronal Cavity

A coronal cavity is a dark zone in the K-corona surrounding prominences. Above cavities arch systems and helmet streamers are often observed. Cavities are interpreted as reduced electron density regions. At the solar limb, in the monochromatic corona, they may take the form of a half-ellipsoid (Figure 6.4).

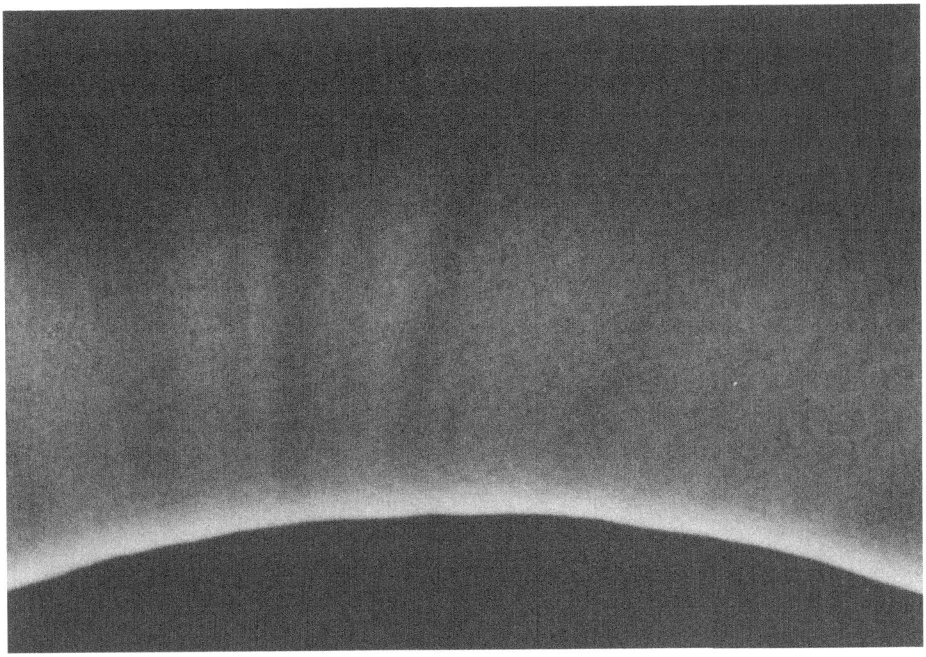

Fig. 6.3. *Polar plumes* observed during the June 30, 1954 solar eclipse. (Courtesy V. Ivanchuk, Kiev Univ.)

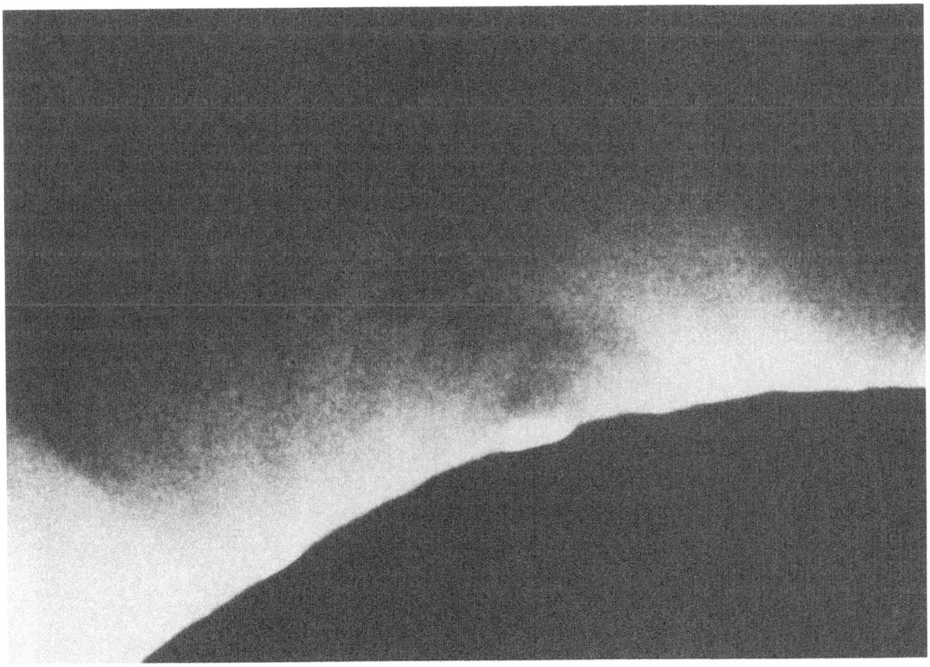

Fig. 6.4. *Coronal cavity* and arch system overlying a prominence observed in the K-corona during the September 22, 1968 eclipse. (Institut d'Astrophysique – CNRS, Paris.)

On the solar disk, *filament cavities* are observed in soft X-ray photographs surrounding the location of quiescent Hα filaments. Filament cavities are closed structures of low emissivity which are not to be confused with coronal holes which are open, presumably low temperature regions. The cavity disappears when the filament disappears, and reappears at about the time of the Hα filament reappearance (Figure 6.9).

It is suggested that cavities are the depletions caused by condensation of material into the prominence/filament.

Saito, K. and Hyder, C. L.: 1968, *Solar Phys.* **5**, 61.
Vaiana, G. S., Davis, J. M., Giacconi, R., Krieger, A. S., Silk, J. K., Timothy, A. F., and Zombeck, M.: 1973. *Astrophys. J.* **185**. L47.
Waldmeier, M.: 1970, *Solar Phys.* **15**, 167.
Webb, D. F.: 1977, *Solar Phys.*, in press.

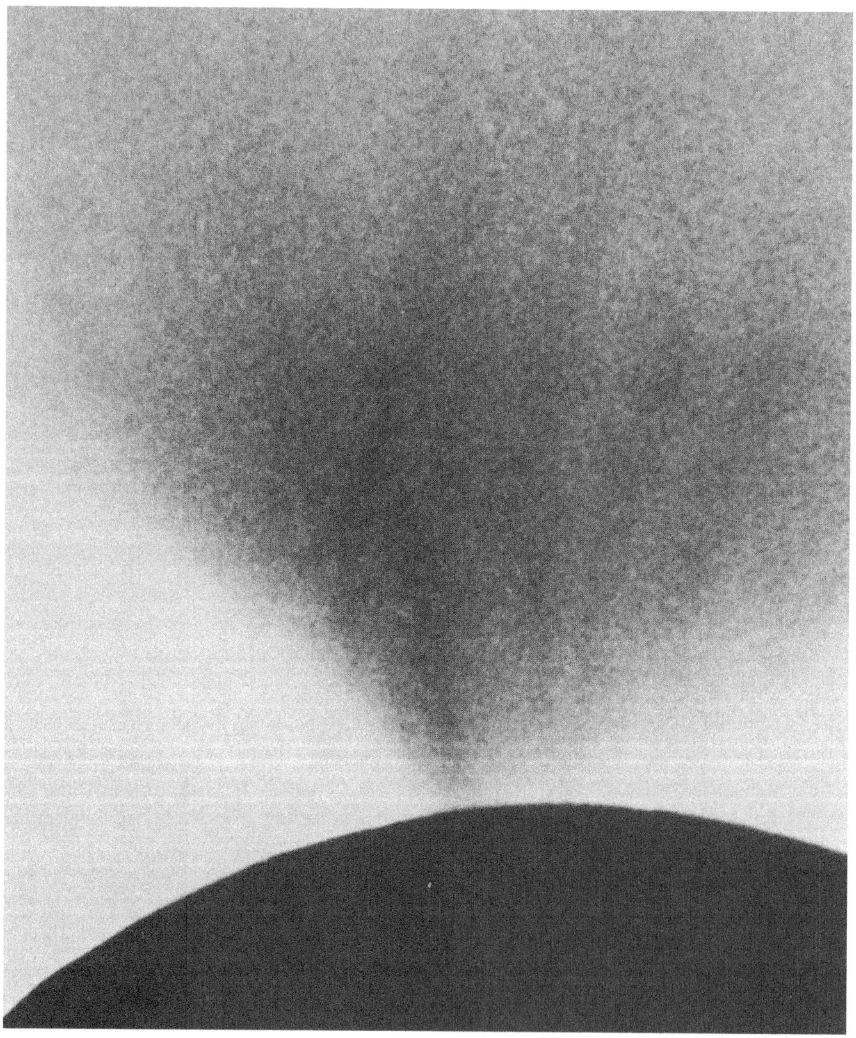

Fig. 6.5. *Rifts* observed at the March 7, 1970 solar eclipse above a coronal hole. (S. Koutchmy.)

6.6 Rifts

Rifts are narrow, dark beams in the K-corona extending sometimes from the solar limb into the outer corona. Often rifts are observed bordering sharp-edge streamers as *dark filaments*, *dark lanes* or *voids*. Rifts should be observed above coronal holes (Figure 6.5).

Koutchmy, S. and Laffineur, M.: 1970. *Nature* **226**, 1141.
MacQueen, R. M., Eddy, J. A., Gosling, J. T., Hildner, E., Munro, R. H., Newkirk, G. A. Jr., Poland, A. I., and Ross, C. L.: 1974, *Astrophys. J.* **187**, L85.

6.7. Coronal Condensation and Enhancement

In the literature there is some confusion between the terms condensation and enhancement. Both terms apply to the coronal parts of active regions, the *active region corona*. Physically, condensation refers to increased (electron) density, enhancement refers to enhanced radiation at visible, EUV, X-ray and radio wavelengths. In the coronal active region one may distinguish a short-lived central condensation proper which is surrounded by a long-lived permanent condensation. Both show complex internal structure (Figure 6.6).

Fig. 6.6. *Coronal condensation* in white light at the September 22, 1968 eclipse. (Courtesy S. Vsekhsvjatsky, Kiev Univ.)

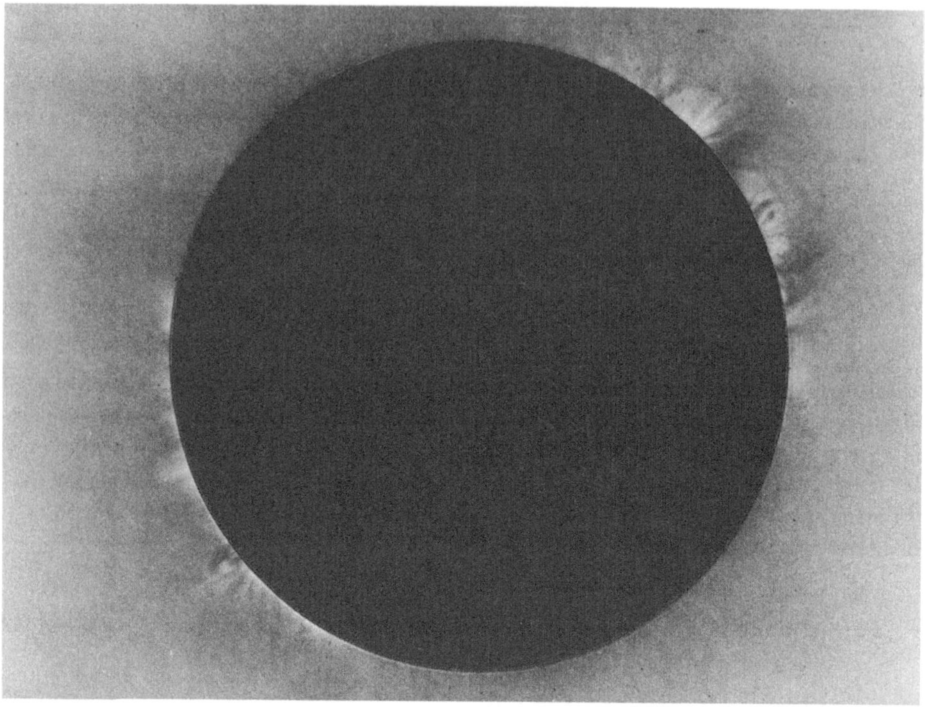

Fig. 6.7. *Monochromatic corona*, green and red coronal line. The superposition of a green line (λ5303 Å) positive with a red line (λ6374 Å) negative shows the structures and temperature inhomogeneities in the lower corona. The regions appearing bright (dominating green line emission) are hotter than the dark ones (dominating red line emission). May 26, 1970. (Courtesy J. L. Leroy, Pic-du-Midi Observatory.)

The *coronal condensation* (*coronal bubble*) is a superdense, hot core (N_e up to 10^{10} cm^{-3}, $T > 3 \times 10^6$ K) with a lifetime of a few days. It may contain a *sporadic condensation* which is flare-associated and has a lifetime of several hours. The sporadic condensation usually consists of a system of bright coronal loops (mainly observed in the monochromatic and soft X-ray corona) which coincide with Hα flare loop prominences (Figure 6.8).

The extended *permanent condensation* has electron densities of a few 10^9 cm^{-3} and temperatures 1.5–2.5 x 10^6 K giving rise to moderately enhanced radiation. Projected against the solar disk the condensations/enhancements coincide with the chromospheric plages and are called then *EUV*, *X-ray* and *radio plages* respectively (see Figure 6.9).

Dunn, R. B.: 1971, in C. J. Macris (ed.), *Physics of the Solar Corona*, D. Reidel Publ. Co., Dordrecht, Holland, p. 114.

Saito, K. and Billings, D. E.: 1964, *Astrophys. J.* **140**, 760.

Waldmeier, M.: 1963, *Z. Astrophys.* **56**, 291.

Zirker, J. B.: 1971, in C. J. Macris (ed.), *Physics of the Solar Corona*, D. Reidel Publ. Co., Dordrecht, Holland, p. 140.

For further references see E-corona.

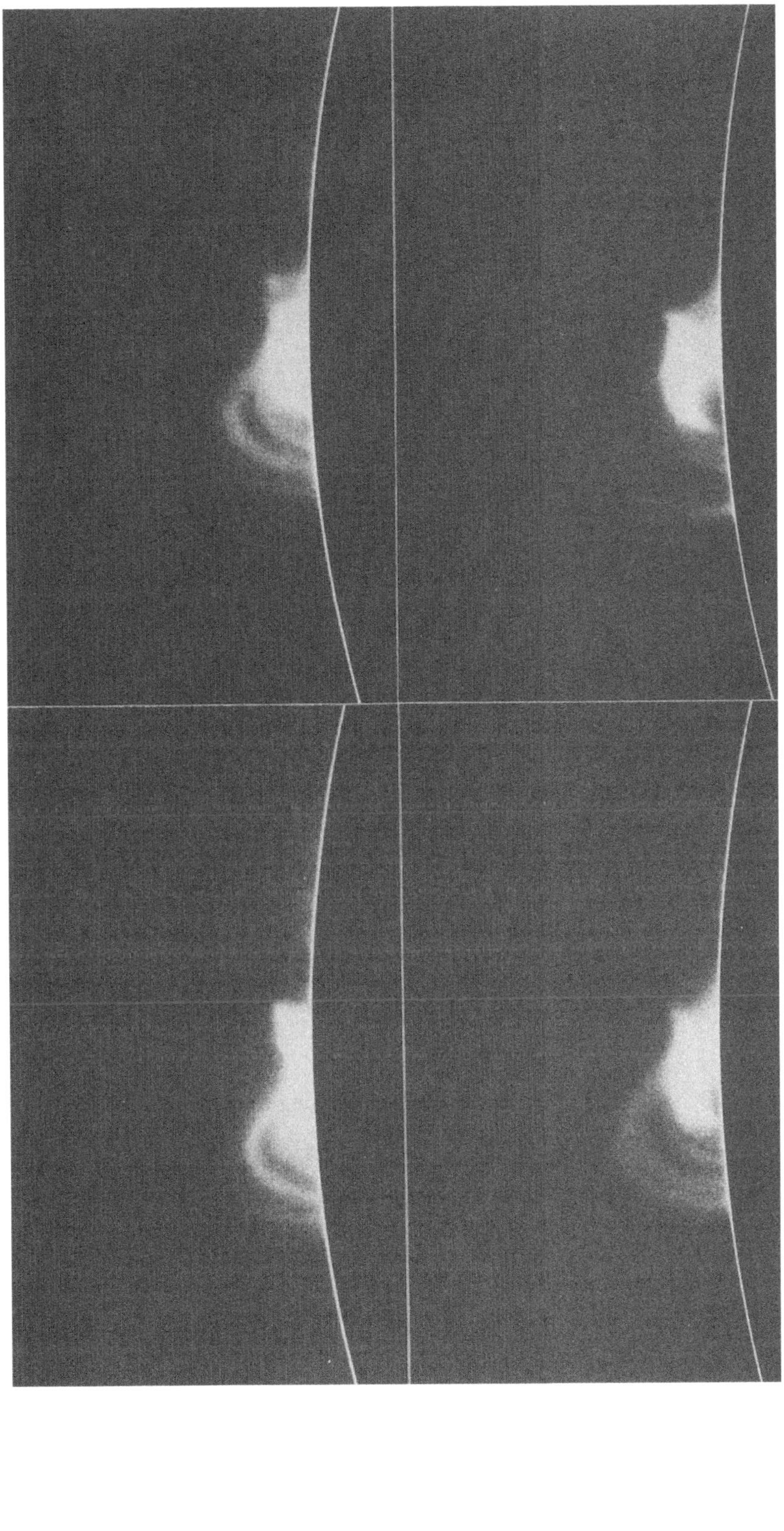

Fig. 6.8. *Sporadic condensation* consisting of tightly packed loops, and expanding *coronal loops* observed in λ5303 Å February 3, 1962. (Sacramento Peak Observatory, Sunspot, N.M.)

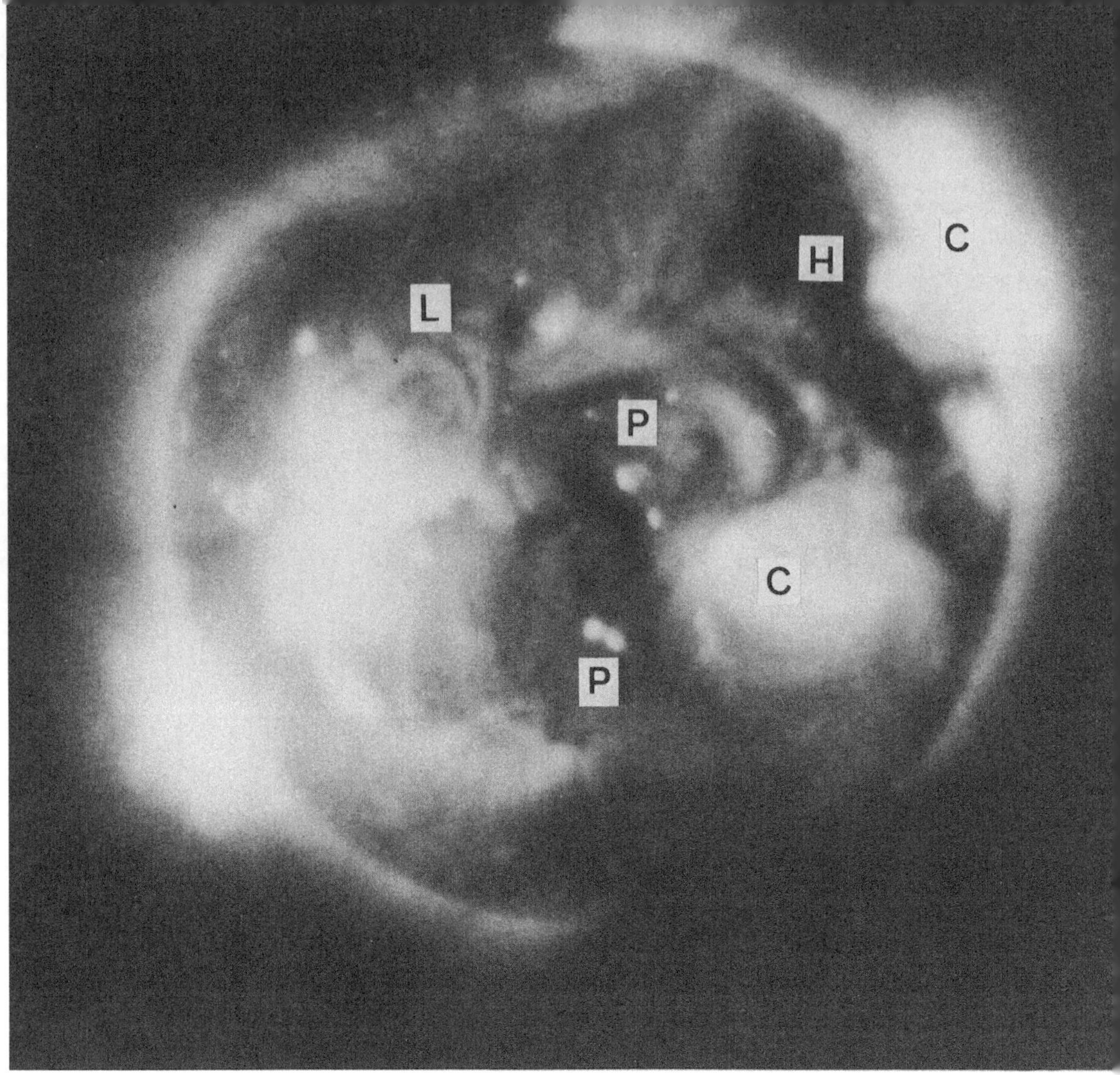

Fig. 6.9. *X-ray corona* photographed in the wavelength bands 3–32 Å and 44–54 Å with the X-ray telescope of American Science and Engineering, Cambridge, Mass. on board Skylab June 23, 1973. It shows coronal holes (*H*), coronal bright points (*P*), coronal condensations (limb and disk) (*C*) and coronal loops (arches) (*L*) against the disk. (Courtesy A. Krieger.)

6.8. E-Corona or Emission Line Corona

Due to its high temperature of the order of a million degrees the solar corona shows line radiation from highly ionized atoms in emission against the continuous spectrum. In the visible spectrum only forbidden lines occur; about 100 lines have been detected during total solar eclipses. The most prominent lines are the red (Fe X λ6374 Å), the green (Fe XIV λ5303 Å) and the yellow (Ca XV λ5694 Å) line. The corona observed in these lines is accordingly called the *green*, *red* and *yellow corona* respectively, or, in general, *monochromatic corona*. It is visible in the inner corona ($< 2\,R_\odot$) only.

Outside total solar eclipses the monochromatic corona is observed employing a Lyot coronagraph with appropriate narrow-band filters or spectrograph. It is strongly heterogeneous being structured into loops, rays and spikes. The structures appear different in different lines because of their different excitation temperatures (Figure 6.7). The local intensity ratio (green line : red line) has been used to derive coronal temperatures.

Dollfus, A.: 1971, in C. J. Macris (ed.), *Physics of the Solar Corona*, D. Reidel Publ. Co., Dordrecht, Holland, p. 97.
Wagner, J. W. and House, L. L.: 1968, *Solar Phys.* **5**, 55.
Waldmeier, M.: 1957, *Die Sonnenkorona*, Birkhäuser, Basel.

6.9. Coronal Loops

Loops are a typical structure of the inner corona observed in the green and red coronal line, in EUV lines and in soft X-rays. They often occur in coronal condensations (enhancements) and are sometimes related to Hα loops (Figure 10.11). They reach heights up to 100×10^3 km and, in general, appear in systems of loops (Figure 6.8, 6.9). The term *coronal arch* usually refers to a large and stable structure in the white-light corona; however in many cases loop and arch are used synonymously.

Dunn, R. B.: 1971, in C. J. Macris (ed.), *Physics of the Solar Corona*, D. Reidel Publ. Co., Dordrecht, p. 114.

6.10. EUV and X-Ray Corona

The solar corona emits radiation in a large number of permitted lines in the EUV and soft X-ray regions and continuous radiation in the X-ray region. Since photospheric radiation is extremely weak at these wavelengths the EUV and X-ray corona is observable also against the solar disk. It shows numerous loops and arches mainly above active regions and connecting different active regions (Figure 6.9). Bright loops observed in extreme-ultraviolet extending outward from strong sources of magnetic flux and having the appearance of a fountain have been called a *magnetic fountain.*

Sheeley, N. R., Jr., Bohlin, J. D., Brueckner, G. E., Purcell, J. D., Scherrer, V., and Tousey, R.: 1975, *Solar Phys.* **40**, 103.
Tousey, R., Bartoe, J.-D. F., Bohlin, J. D., Brueckner, G. E., Purcell, J. D., Scherrer, V. E., Sheeley, Jr., N. R., Schumacher, R. J., and Van Hoosier M. E.: 1973, *Solar Phys.* **33**, 265.
Vaiana, G. S., Krieger, A. S., and Timothy, A. F.: 1973, *Solar Phys.* **32**, 81.

6.11. Magnetic Arcades

When patterns of coronal fieldlines are computed from measured photospheric fields, series of magnetic loops are found which form corridors or arcades extending over more than ¼ of the solar circumference. While *Low Magnetic Arcades* (LMA) ($h_{max} < 1.5\ R_\odot$) may have any orientation. *High Magnetic Arcades* (HMA) ($h_{max} > 1.5\ R_\odot$) show a decided preference for east-west orientation. The highest loops of HMA in the current-free approximation are drawn out into forms reminiscent of coronal streamers.

In the two-dimensional projection in the plane of the sky arcades appear as *magnetic arches*; they coincide with closed white-light coronal arches associated with the bases of coronal streamers. Arcades have been found also over solar sector-structure boundaries.

X-ray photographs of the corona against the disk show *active region interconnections* (large scale arches connecting active regions) which probably represent HMA.

Newkirk, G., Jr. and Altschuler, M. D.: 1970, *Solar Phys.* **13**, 131.
Vaiana, G. S., Krieger, A. S., and Timothy, A. F.: 1973, *Solar Phys.* **32**, 81.
Wilcox, J. M. and Svalgaard, L.: 1974, *Solar Phys.* **34**, 461.

6.12. Coronal Holes

These are extended regions of exceptionally low density and temperature in the solar corona. They appear deficient in coronal EUV, X-ray and cm radiations but are not observable in spectral lines produced at $T \leqslant 600\,000$ K (Figure 6.9). Holes are probably associated with unipolar regions whose photospheric magnetic fields diverge in the corona. They appear to be sources of strong solar wind. The lifetime of holes is typically of the order of several solar rotations and they rotate like a rigid body.

The metric radio emission from coronal holes is similar to that of polar regions, the minimum corona and the quiet radio Sun.

Fürst, E. and Hirth, W.: 1975, *Solar Phys.* **42**, 157.
Nolte, J. T., Krieger, A. S., Timothy, A. F., Vaiana, G. S., and Zombeck, M. V.: 1976, *Solar Phys.* **46**, 291.
Nolte, J. T., Krieger, A. S., Timothy, A. F., Gold, R. E., Roelof, E. C., Vaiana, G., Lazarus, A. J., Sullivan, J. D., McIntosh, P. S.: 1976, *Solar Phys.* **46**, 303.
Timothy, A. F., Krieger, A. S., and Vaiana, G. S.: 1975, *Solar Phys.* **42**, 135.

6.13. Coronal Bright Points

These are observed in X-ray telescope images to have a typical diameter of 30″ and a bright core of 10″. They have a mean lifetime of 8 hr and are uniformly distributed across the solar surface. Coronal bright points are associated with small bipolar photospheric magnetic regions (estimated field strength ≈ 10 G). About 1500 X-ray points emerge per day, possibly bringing more new magnetic flux to the surface than active regions. Bright points represent a type of solar activity different from normal active regions and ephemeral regions although an association with the latter is suggested (Figure 6.9).

Golub, L., Krieger, A. S., Silk, J. K., Timothy, A. F., and Vaiana, G. S.: 1974, *Astrophys. J.* **189**, L93.

6.14. Coronal Events or Coronal Transients

These are general terms for a multitude of changes in the corona such as suddenly brightening or fading structures, expanding, ascending or disrupting arches, moving clouds etc. Such features have been observed in the monochromatic and X-ray corona as well as in the inner and outer K-corona. Many of them are associated with flares and/or with ascending or eruptive prominences and with type II and type IV radiobursts.

The *monochromatic transients* or events have been classified into slow events, loop and arch events and fast events. Types of *fast events* are: oscillations of structures, disruption of arches, realignments of arch systems, accelerated expansion of arches (*coronal whip*).

Four basic types of *inner K-corona transients* have been observed with coronameters: formation of a streamer or condensation, disruption of a streamer, relocation of a structure and the expansion of an arch. A special type is the *abrupt depletion* – a sharp drop of electron density seemingly due to expulsion of plasma into the outer corona.

With satellite-borne coronagraphs *outer corona transients* such as disintegration of bright streamers and arches and outward moving plasma clouds ($v \approx 1000$ km s^{-1}) as well as apparent shock phenomena have been observed (Figure 6.10) at an average rate of one every 40 hours; at periods of high solar activity they are expected to occur one every 10 hr. About two thirds of the brightest outer coronal transients observed by Skylab were associated with erupting prominences. The observations suggest, however, that the

Fig. 6.10. *Coronal transient* associated with a prominence eruption observed by the HAO coronagraph aboard Skylab June 10, 1973, 0943 UT. The loop structure is material moving outward with an apparent velocity of approximately 450 km s^{-1}. The event was observed for about half an hour. (Courtesy R. MacQueen.)

material of the transient comes from the lower corona above the prominence rather than from the prominence itself.

Dunn, R. B.: 1971, in C. J. Macris (ed.), *Physics of the Solar Corona*, D. Reidel Publ. Co., Dordrecht, p. 114.
Hansen, R. T., Garcia, C. J., Hansen, S. F., and Yasukaura, E.: 1974, *Publ. Astron. Soc. Pacific* **86**, 500.
Hildner, E., Gosling, J. T., Hansen, R. T., and Bohlin, J. D.: 1975, *Solar Phys.* **45**, 363.
Hundhausen, A. J. and Newkirk, G., Jr. (eds.): 1974, *Flare-Produced Shock Waves in the Corona and in Interplanetary Space*, High Altitude Observatory, Boulder.
Newkirk, G., Jr. (ed.): 1974, 'Coronal Disturbances', *IAU Symp.* **57**.

6.15. T-Corona

The T-corona, *local F-corona* or *circum-solar dust emission* is thermal emission from interplanetary or circum-solar dust re-emitting absorbed solar radiation at infrared wavelengths. It has been observed at $\lambda 2.2\mu$ at distances of $3.5-10\, R_\odot$ as an enhancement concentrated in the equatorial plane. A solar origin for the interplanetary dust grains has been suggested.

MacQueen, R. M.: 1968, *Astrophys. J.* **154**, 1059.

6.16. Lyot-Coronagraph

This is a telescope designed for the investigation of the inner corona outside solar eclipses. The Sun is occulted by a disk of appropriate size placed in the primary focus of a single objective lens (internal occultation). The Lyot-coronagraph is used with a filter or spectrograph for routine observations of the monochromatic corona.

Combined with a polarimeter it serves as a *K*-coronameter which determines – outside eclipses – the intensity B of the K-corona multiplied by p the degree of polarization (*pB-corona*). The polarimeter separates the faint, strongly polarized radiation of the K-corona from the superposed strong, but unpolarized, atmospheric stray light.

Dollfus, A.: 1974, in T. Gehrels (ed.), *Planets, Stars and Nebulae Studied with Photopolarimetry*, Univ. Arizona Press, Tucson, p. 695.
Hansen, R. T., Hansen, S. F., and Price, S.: 1966, *Publ. Astron. Soc. Pacific* **78**, 14.
Wlérick, G. and Axtell, J.: 1957, *Astrophys. J.* **126**, 253.

6.17. Externally Occulted Coronagraph

This type of coronagraph is used with balloon, rocket and satellite experiments. The Sun is occulted by a disk placed some distance in front of the objective lens. The inner corona is vignetted but the outer parts of the white-light corona beyond $2\, R_\odot$ can be observed (Figure 6.10).

Koomen, M. J., Detwiler, C. R., Brueckner, G. E., Cooper, H. W., and Tousey, R.: 1975, *Appl. Opt.* **14**, 743.
Newkirk, G., Jr. and Bohlin, J. D.: 1963, *Appl. Opt.* **2**, 131.

7. ACTIVE REGIONS

M. J. MARTRES and A. BRUZEK

7.1. (Solar) Activity

The terms 'active' and 'activity' are used in solar physics in a rather general and not well-defined way. Basically, 'activity' denotes a marked degree of variability in time associated with enhanced motions, density, temperature, emission etc. It denotes further the occurrence, or frequent occurrence of a certain (transient) phenomenon (e.g. flare activity, surge activity, X-ray activity) or the occurrence of 'active regions' ('solar activity' in general). Radio activity, for example, can imply both the occurrence of radio bursts and the longer-lasting enhancement of radio emission.

Solar features and regions are qualified as 'active' if they show activity as described above; active prominences, for instance, show enhanced internal or large scale motions, changes in shape, increase in brightness etc. For 'active region' see next entry.

7.2. Active Region

The solar Active Region (AR) is an extremely complex phenomenon comprising a large variety of features (active region phenomena) in the photosphere, chromosphere and corona. The occurrence of the various active phenomena depends on the phase and state of evolution of the AR; their appearance depends on the radiation used for the observation (cf. Figures 3.1, 4.4, 6.9, 7.1, 8.1). There is a large variation in size, appearance and lifetime among different AR's. Lifetimes range from several hours up to a few months. Size and appearance depend largely on the age of the AR.

Identity is given to an AR by a strong magnetic field which controls its development and character. Subsurface magnetic flux emerges locally to form an extended bipolar or multipolar pattern with strong concentration in sunspots. The total magnetic flux in a medium-sized AR is $\approx 10^{22}$ Mx. Physically, AR's outside sunspots are characterized by enhanced density and temperature and accordingly enhanced radiation throughout the spectrum from X-rays to radio waves (except some line emission). For instance, the density in the coronal part of an AR (the coronal condensation) may be increased 10 times, its temperature 5 times and the radio emission by several orders of magnitude. Characteristic for AR's are also large-scale plasma flows (v up to ≈ 100 km s^{-1}) and the acceleration of particles. Enhanced radiation in the EUV to X-ray range as well as energetic particles from AR's affect the interplanetary medium and the terrestrial magnetosphere and upper atmosphere (ionosphere).

Bruzek, A.: 1972, in E. R. Dyer (ed.), *Solar Terrestrial Physics*, 1970, D. Reidel Publ. Co., Dordrecht, Part I, p. 49.
Kiepenheuer, K. O. (ed.).: 1968, 'Structure and Development of Solar Active Regions', *IAU Symp.* **35**.
Tandberg-Hanssen, E.: 1967, *Solar Activity*, Blaisdell Publ. Co., Waltham, Mass.

Bruzek and Durrant (eds.), Illustrated Glossary for Solar and Solar-Terrestrial Physics. 53–70.

7.3. Complexes of Activity

New active regions tend to develop close to existing regions or at the approximate (Carrington) longitude of decayed regions. Thus active regions cluster and form long sequences at *preferred* or *active longitudes* forming *complexes of activity* or *families*. Complexes appear to rotate at a fixed period of 27 days independent of solar latitude.

Preferred longitudes have been observed to persist for several years. They have been detected also in the occurrence of proton flares. It seems that pairs of preferred longitudes separated by 180° exist. A possible interpretation of this clustering of solar activity is that the regions of a complex emerge from a persistent subsurface source rotating with a fixed period.

French authors prefer the term *centre of activity* (CA) for the single active region and use the term active region for a complex of activity (composed of several CA's).

See Figure 7.1a.

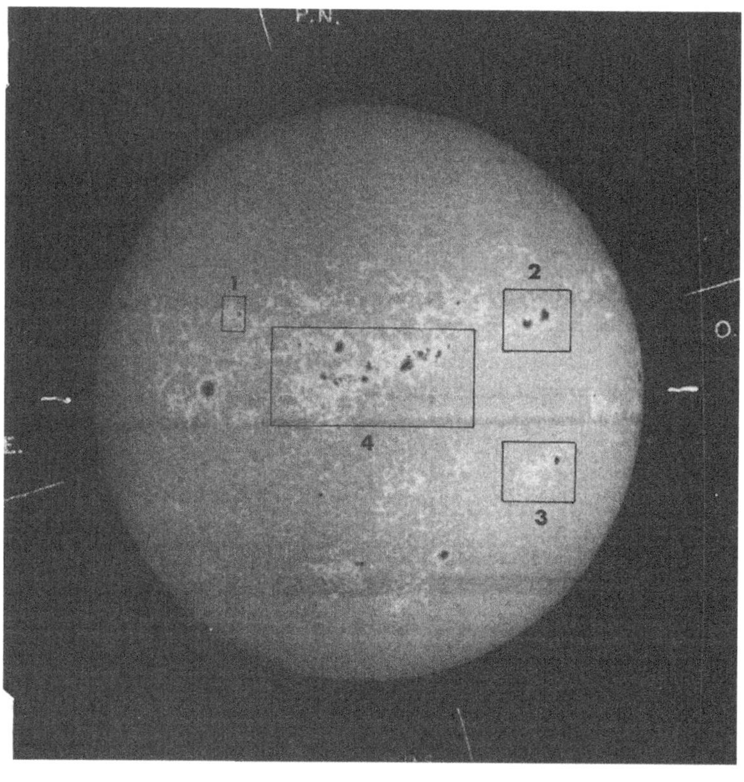

Fig. 7.1. The Sun with active regions observed in various wavelengths.

Fig. 7.1a. Spectroheliogram taken in K_{1v} i.e. in the violet wing of the Ca II K-line λ3967 Å. (1), (2), (3) are isolated active regions in different phases of evolution containing spot groups; (4) is a complex of activity. Other plage regions with and without spots are present (photograph: Observatoire de Meudon, August 13, 1974). For active regions photographed in the K-line centre (Ca-plages) see Figure 4.4. for photospheric faculae (white-light photograph) Figure 8.1.

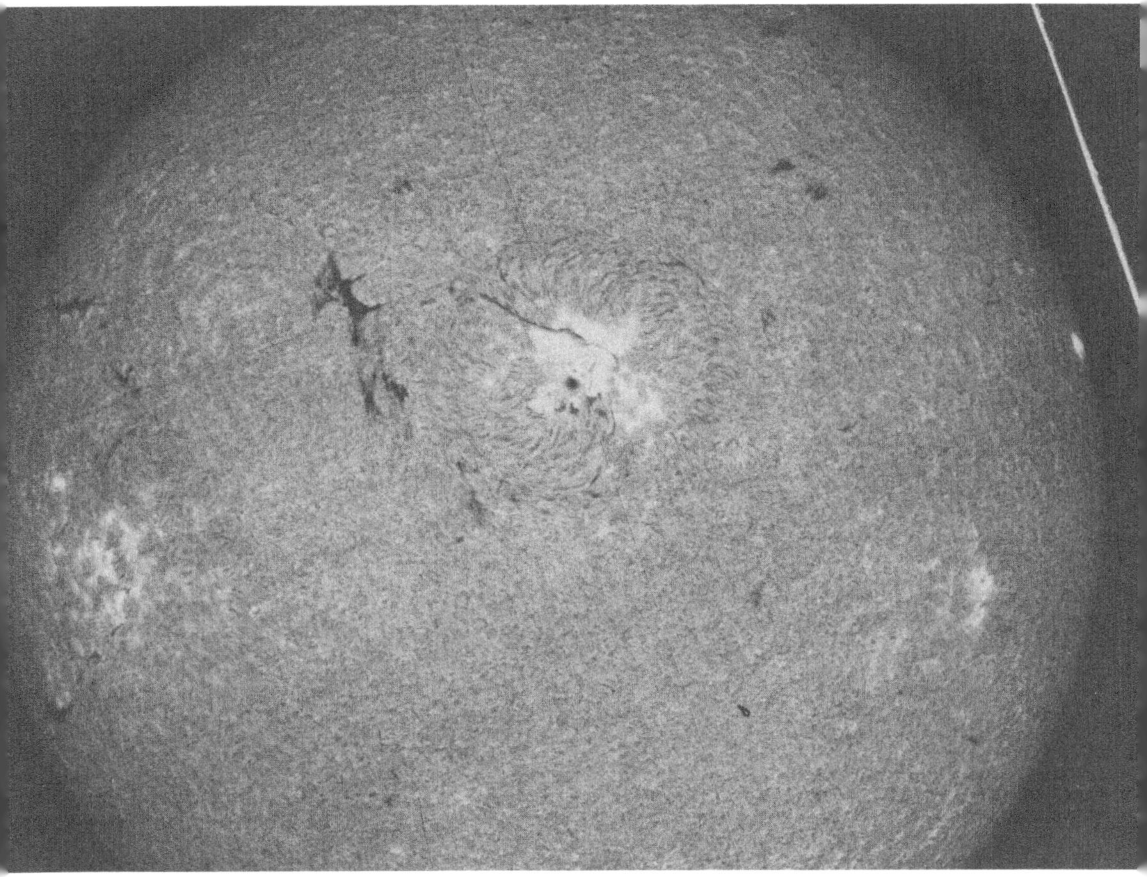

Fig. 7.1b. Spectroheliogram taken in Hα (hydrogen λ6563 Å); it shows: active region near the disk centre surrounded by a large solar vortex; small young AR's with bright plage (EFR) near both limbs; an old scattered plage near the E-limb (left); a number of dark filaments. (Photograph: Observatoire de Meudon, May 21, 1973.)

Becker, U.: 1955, *Z. Astrophys.* **37**, 47.

Bumba, V. and Howard, R.: 1965, *Astrophys. J.* **141**, 1492, 1502.

Dodson, H. W. and Hedeman, E. R.: in Kiepenheuer (ed.), 'Structure and Development of Solar Active Regions', *IAU Symp.* **35**, 56.

Haurwitz, M.: 1968, *Astrophys. J.* **151**, 351.

7.4. Emerging Flux Regions

Emerging Flux Regions (EFR's) form the first stage of active regions although EFR's also appear within and at the borders of existing active regions; they bring new magnetic flux to the surface of the Sun. The EFR is a bipolar magnetic region which first produces a small bipolar plage in Hα and K connected by a few tiny dark filaments (Figure 7.2). About one day later a *Bright Region with Loops* (BRL) is formed, that is an enlarged bright plage crossed by a system of dark loops known as an *Arch Filament System* (AFS). The AR contains at this stage a bipolar spot group with the spots at the opposite ends of the arch filaments.

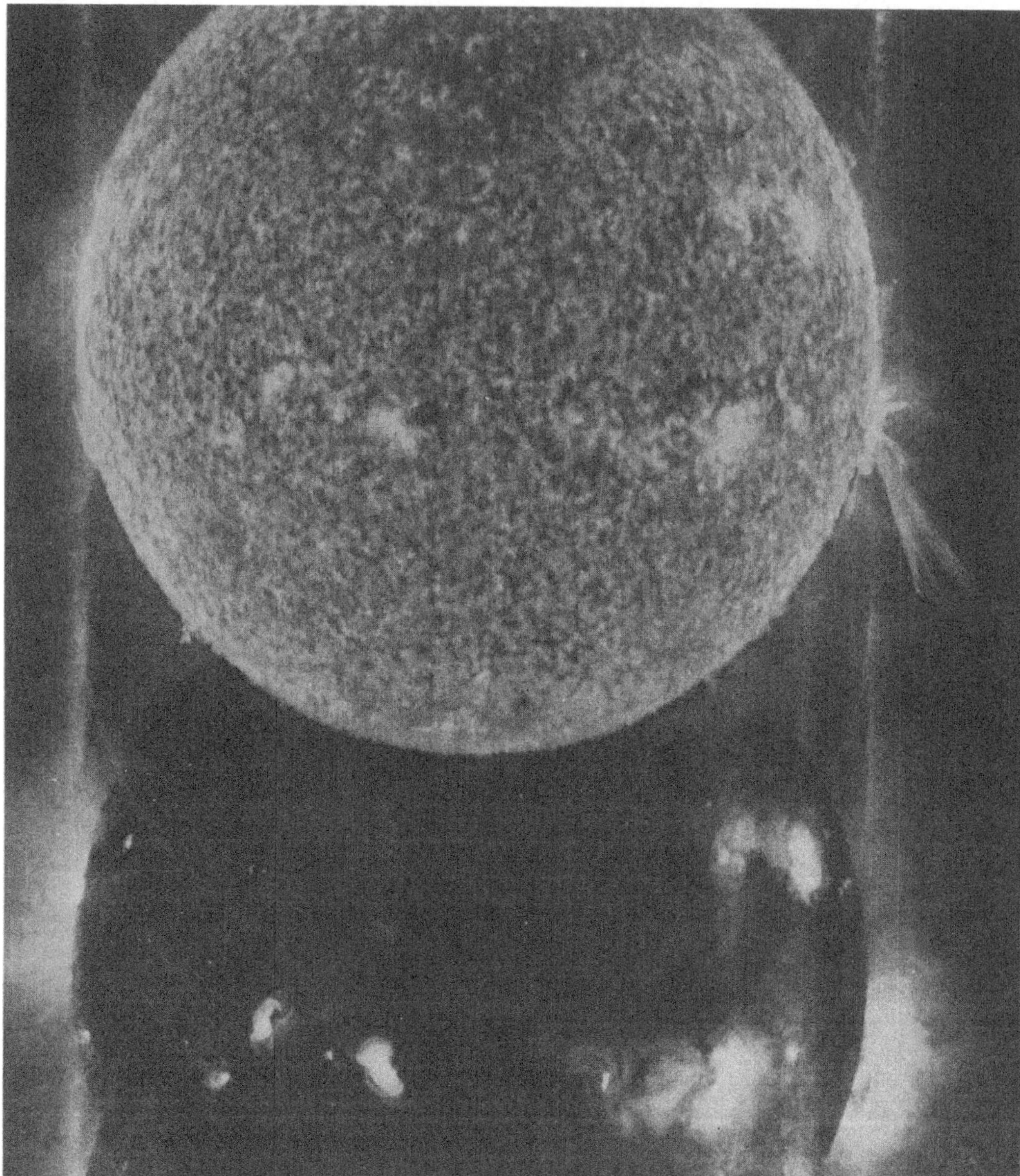

Fig. 7.1c. Photographs taken in the He II λ304 Å line (top) and in the Fe XV λ284 Å line showing active regions on the solar disk as they appear in the chromosphere and in the EUV corona. In the coronal line a large coronal enhancement is visible above the right-hand solar limb; in the He II line there is a large active prominence. (Naval Research Laboratory photograph.) For AR's in (coronal) X-rays see Figure 6.9.

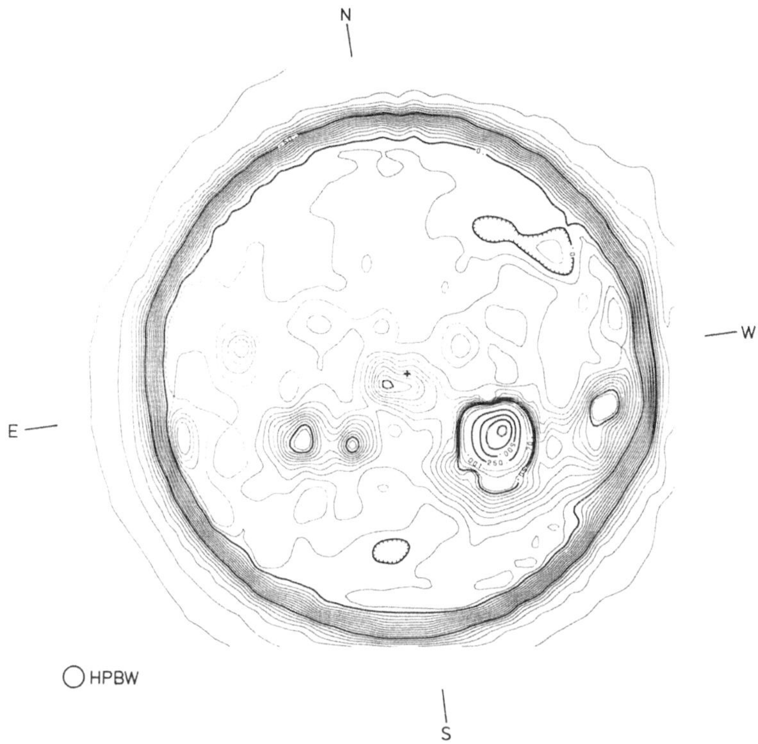

Fig. 7.1d. Radio Sun observed at λ 3 cm showing AR's as regions of enhanced radio emission (radio plages). (Courtesy O. Hachenberg, Max Planck Institut f. Radioastronomie, Bonn.) Active regions observed as magnetic regions see Figure 3.1.

The arch filaments are flat arches (length up to 30 000 km, height < 5000 km) which are believed to outline magnetic field loops (Figure 7.3). The centres (the tops) of the arches rise at $\lesssim 10$ km s^{-1} while material descends near both ends at $v \lesssim 50$ km s^{-1}. This suggests that magnetic flux loops are expanding through the chromosphere into the inner corona. Individual arches fade after ≈ 20 min and are replaced by another generation from below.

This phase of an EFR (and AR) with conspicuous AFS may last 3–4 days until a well developed bipolar group (type D) is formed.

Bruzek, A.: 1969, *Solar Phys.* **8**, 29.
Frazier, E. N.: 1972, *Solar Phys.* **24**, 98.
Zirin, H.: 1974, in R. G. Athay (ed.), 'Chromospheric Fine Structure', *IAU Symp.* **56**, 161.

7.5. Ephemeral Region

This term refers to short-lived, small emerging flux regions. They have been recognized on K and Hα spectroheliograms but are especially prominent on magnetograms as small bipolar magnetic features. Ephemeral regions appear to be simply the small-scale end of a

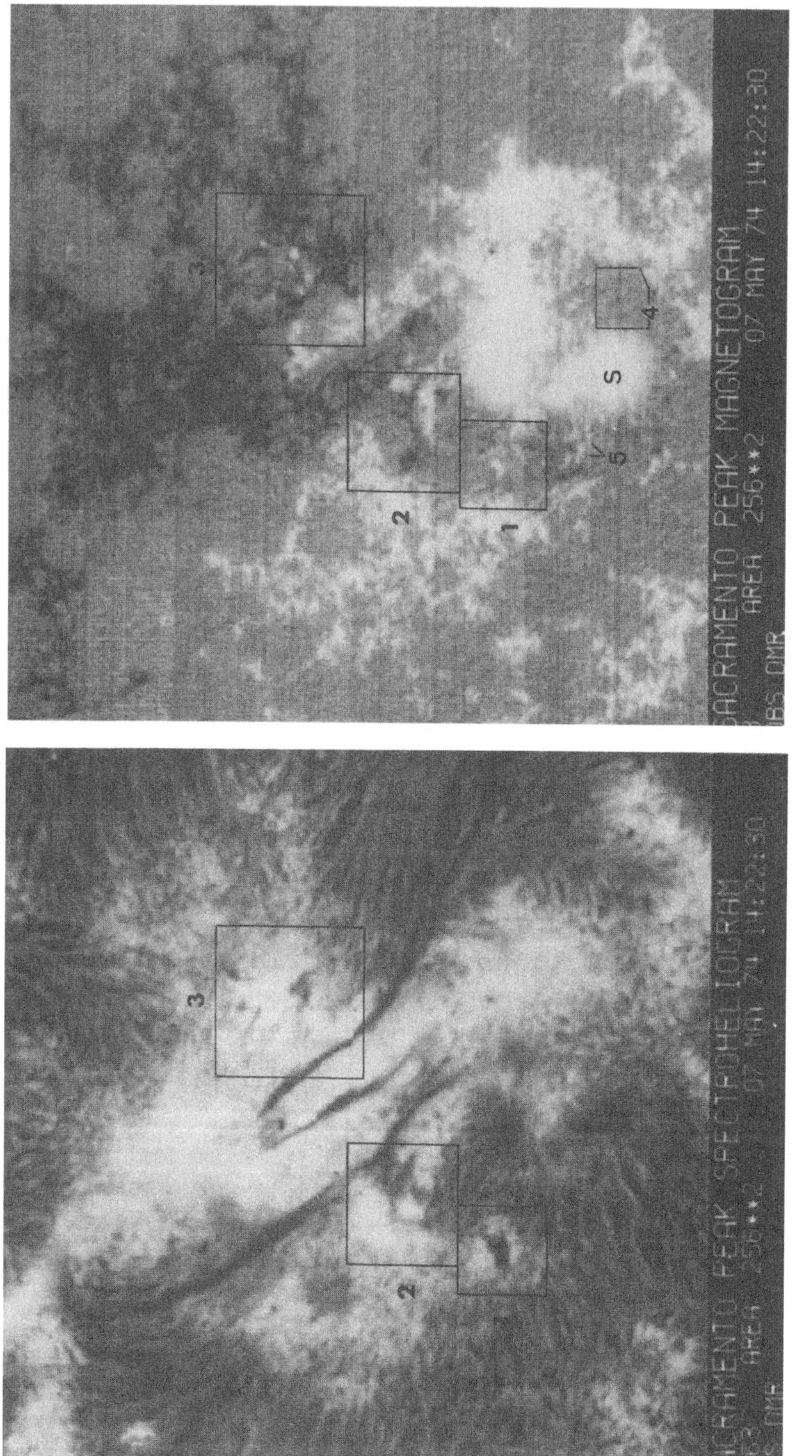

Fig. 7.2. Sacramento Peak Hα filtergram and magnetogram showing three emerging flux regions (1, 2, 3) in order of growing importance. Also shown are (4) a moat and (5) moving magnetic features (S spot). Note moreover the dark Hα filaments indicating the inversion line between magnetic field regions of opposite polarity. The area covered is 256 by 256″. (Courtesy D. M. Rust.)

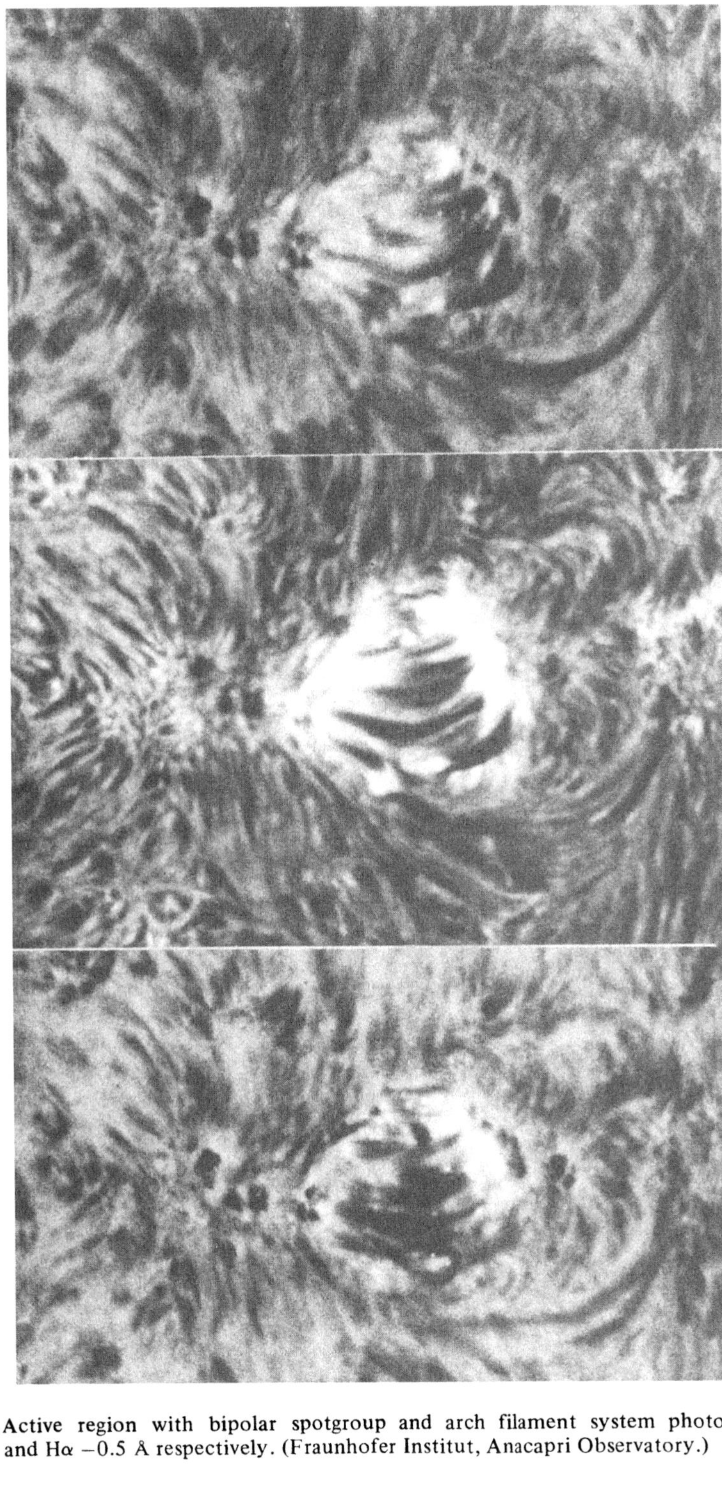

Fig. 7.3. Active region with bipolar spotgroup and arch filament system photographed in Hα +0.5 Å, Hα and Hα −0.5 Å respectively. (Fraunhofer Institut, Anacapri Observatory.)

broad spectrum of active region sizes; they develop no spots. It is possible that some fraction of the ephemeral regions are due to processes fundamentally different from those that give rise to regular active regions. Ephemeral regions are characterized by areas of less than 100 millionths of the visible hemisphere, lifetimes of ≈ 1 day, total magnetic flux of $\approx 10^{20}$ Mx, a broad latitude and longitude distribution and a numerical variation with the solar cycle. About 100 ephemeral regions may be formed during one day (Figure 7.4.).

Related to ephemeral regions are X-ray and EUV coronal bright points but not all ephemeral regions recognized on magnetograms are associated with bright points. Ephemeral regions appear to be the source of as much magnetic flux as larger active regions and may be significant in heating the atmosphere.

Harvey, K. L., Harvey, J. W., and Martin, S. F.: 1975, *Solar Phys.* **40**, 87.

7.6. Sunspot Group

Sunspots are concentrations of magnetic flux which are arranged in the centre of an AR

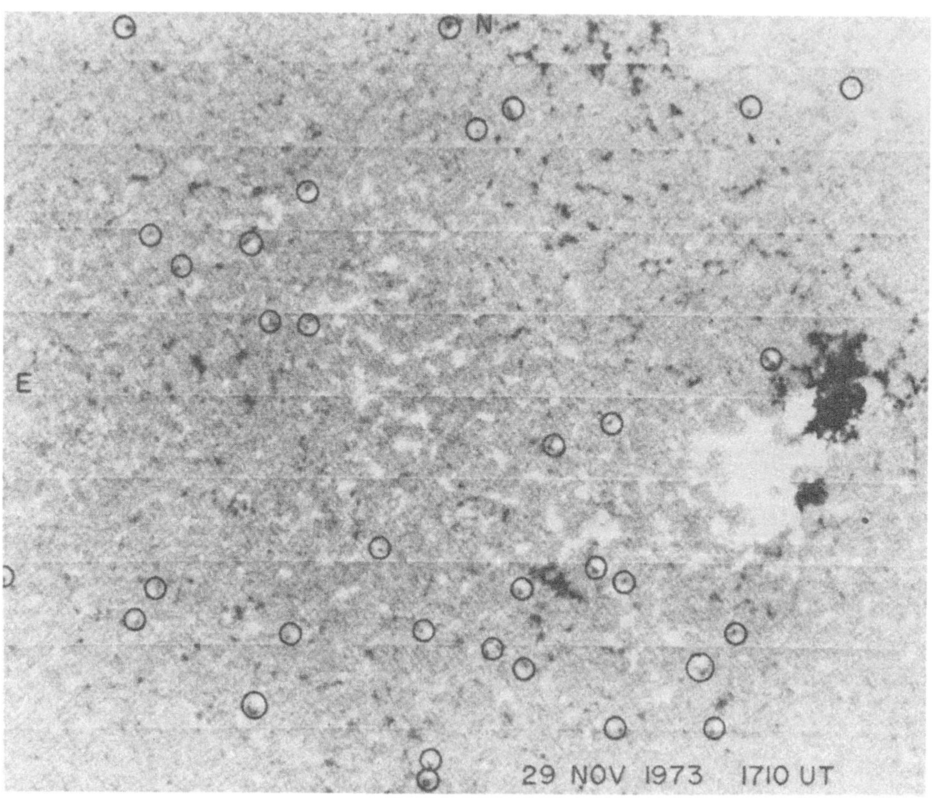

Fig. 7.4. Magnetogram showing ephemeral regions (circled); bright and dark regions indicate fields of opposite polarity respectively. (Harvey *et al.*: 1975, *Solar Phys.* **40**, 87, Figure 1.)

(plage) as an elongated, magnetically bipolar or multipolar spot group. The western part of a spot group is called the preceding part; the main spot in it, usually a large, rather regular spot is called the *p-spot* or the *leader* (*spot*) of the group. The main spot of the easterly, the following part is called the *f-spot* or *follower* (cf. Figure 7.1a).

The development and decay of spot groups follows a pattern which is schematically outlined by the sequence A to J in the *Zürich Spot Classification*: Emerging magnetic flux gives rise first to a small cluster (Zürich type A) or a bipolar group (type B) of several small spots or pores. This becomes a group of well developed spots (with penumbra) stretching over 5 to 10 heliocentric degrees (types C and D) within 2–4 days. Additional emerging flux makes the group complex with large spots and an extension of up to 15–20 heliocentric degrees within another 3–6 days (types E and F). Maximum development is usually reached 8–10 days after birth with a total spot area of up to several thousandths of the area of the visible hemisphere; the maximum rate of growth may be $\approx 200 \times 10^{-6}$ of the visible hemisphere per day. The growth may, however, stop and decay may start at any of the above stages (types). Decay occurs initially at the same rate as the growth, but leaves the leader spot almost unaffected (type G), so that after 4–10 days – depending on the maximum type reached – the leader only is left with a few companions (types H and J). The leader decays at a slow constant rate of 6×10^{-6} per day (independent of the actual size of the spot!). This last phase of spot decay may therefore last one or more solar rotations until the active region is spotless (Figure 7.5).

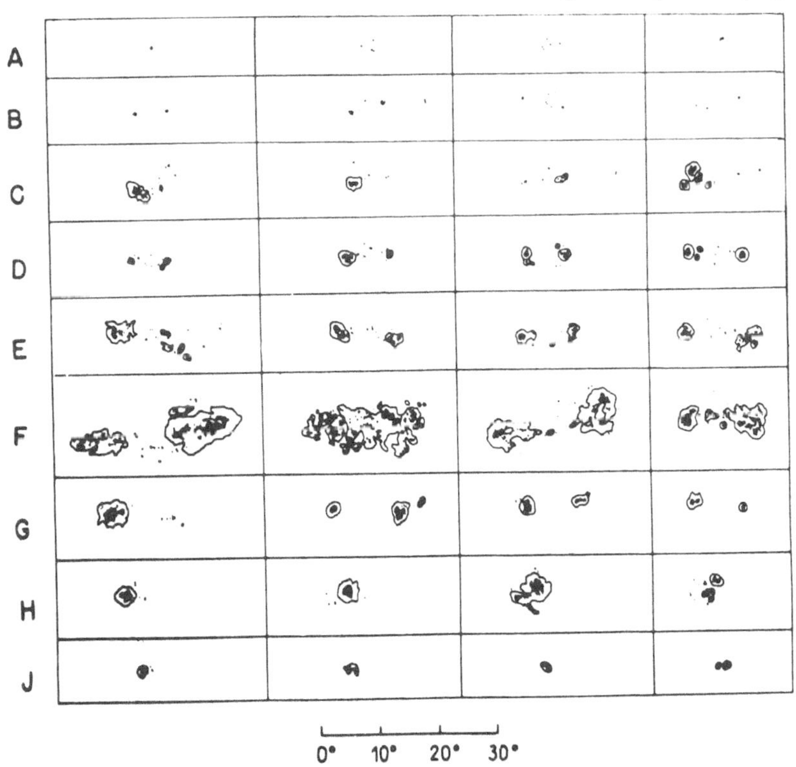

Fig. 7.5. Zürich classification of sunspot groups. (Waldmeier, M.: 1947, *Publ. Sternwarte Zürich* 9, 2.)

The AR is most (surge-, flare-, radio) active during rapid spot growth and changing spot configuration, to a lesser degree during the first decay phase (type G) and is virtually inactive for the H and J type phases unless there is some revival of activity brought about by newly emerging flux.

The *Solar Geophysical Data* uses a modified Zürich Classification which has been extended by McIntosh, adding information on the type of penumbra and on the spot distribution within the group.

de Jager, C.: 1959, in S. Flügge (ed.), *The Solar System*, (Encyclopedia of Physics, Vol. LII), Springer Verlag, Berlin, p. 166.
Kiepenheuer, K. O.: 1953, in G. P. Kuiper (ed.), *The Sun*, Univ. of Chicago Press, Chicago, p. 340.
Solar Geophysical Data: 1975, No. 366 Supplement, p. 30.

7.7. Proper Motion (of Sunspots)

There are different types of proper motions of sunspots, i.e. motions relative to the surrounding photosphere:

(1) During the growth of spot groups there is a relative extension both in longitude and, to a lesser degree, in latitude. The proper motion in longitude is largest for the *p*-spot during the first few days of group development (≈0.6 heliocentric degrees per day to the West). It reverses its direction when the maximum size of the group is attained, and the *p*-spot eventually returns to approximately its original position in longitude. The motion in latitude is largest during the reversal of the motion in longitude and is usually $\lesssim 0^\circ 3$/day. The proper motions of the *f*-spot are much smaller.

(2) Spot umbra may divide and eject a 'daughter umbra' which moves away at $\approx 1^\circ$/day. The reverse process of merging of spots has also been observed.

(3) Another type is the interpenetration of two spot groups or the apparent passage of a single spot through a spot group.

(4) A special type is the rotational motion of a single spot or of a spot pair.

Types 2 and 4 have been observed in spot groups which produced series of very large flares.

Bruzek, A.: 1960, *Z. Astrophys.* **50**, 110.
Kiepenheuer, K. O.: 1953, in G. P. Kuiper (ed.), *The Sun*, Univ. of Chicago Press, Chicago, p. 166.

7.8. Evolving Magnetic Features (EMF)

The photospheric longitudinal magnetic fields of active regions can be divided into a number of local peaks separated by zones of weaker fields. These unipolar features may include a spot or a cluster of spots, or may contain no spot at all. During the life of an AR they appear, grow, and die, thus determining the magnetic configuration and development of the AR. The field intensity and size of these evolving magnetic features vary greatly: the magnetic field strengths range between 100 and 2000 G; characteristic dimensions are 10^4-10^5 km. Flare occurrence is found to be associated with complementary changes of magnetic flux in adjacent EMF's of opposite polarity.

Martres, M.-J., Michard, R., Soru-Iscovici, I., and Tsap, T. T.: 1968, *Solar Phys.* **5**, 187.
Ribes, E.: 1969, *Astron. Astrophys.* **2**, 316.

7.9. Parasites or Inclusions

These are isolated, rather strong magnetic field peaks in AR's which are completely surrounded by areas of opposite polarity. They contribute to the complexity of the field configuration and appear to be preferred places of flare origin.

Forming a special type of this magnetic feature are the *satellites* or *satellite spots*: large spots are frequently surrounded by a number of small magnetic peaks with a polarity opposite to that of the central spot. In most cases these magnetic satellites are not visible as spots in white light. They have been called *invisible spots* – a general term for strong magnetic fields peaks not producing a visible spot. This misnomer arose before the realisation that non-spot fields of surrounding plage are also concentrated in small-scale structures.

Satellites appear to be closely associated with the production of flares, surges and moustaches. They must not be confused with the moving magnetic features and magnetic knots both of which are short lived and very small features.

Martres, M. J., Michard, R., and Soru-Iscovici, I.: 1966, *Ann. Astrophys.* **29**, 245.
Rust, D. M.: 1968 in K. O. Kiepenheuer (ed.), 'Structure and Development of Solar Active Regions', *IAU Symp.* **35**, 77.
Severny, A. B.: 1964, *Space Sci. Rev.* **3**, 451.

7.10. Magnetic Inversion Line

The line which separates longitudinal magnetic fields of opposite polarity has been given several names by various authors: *dividing line*, *transition line* or *line of inversion* are correct terms. *Zero line* and the frequently used *neutral line* are misleading and ought to be avoided; they imply that there is no field at all whereas, in general, there is a transverse field, in many cases a strong one.

The inversion line is often superposed by a dark filament which, in many cases inside AR's, is rather thin and would not appear as a prominence at the solar limb. Filaments can therefore be used, with some caution, as a tracer of inversion lines (see Figure 7.2). The areas close to and along inversion lines are preferred places of flare occurrence.

7.11. Filament Channel or Plage Couloir

In K_1 and in the Hα wing, the base of a filament appears replaced by a couloir or channel between opposite magnetic polarities. This channel may even exist without a filament along the inversion line. It is completely void containing neither plage nor chromospheric fine structure (Figure 7.6).

The size of the filament and the width of the channel are inversely proportional to the field gradient across the magnetic inversion line. The filament is very thin or even absent and no channel is visible if the polarity changes abruptly as in young AR's or between

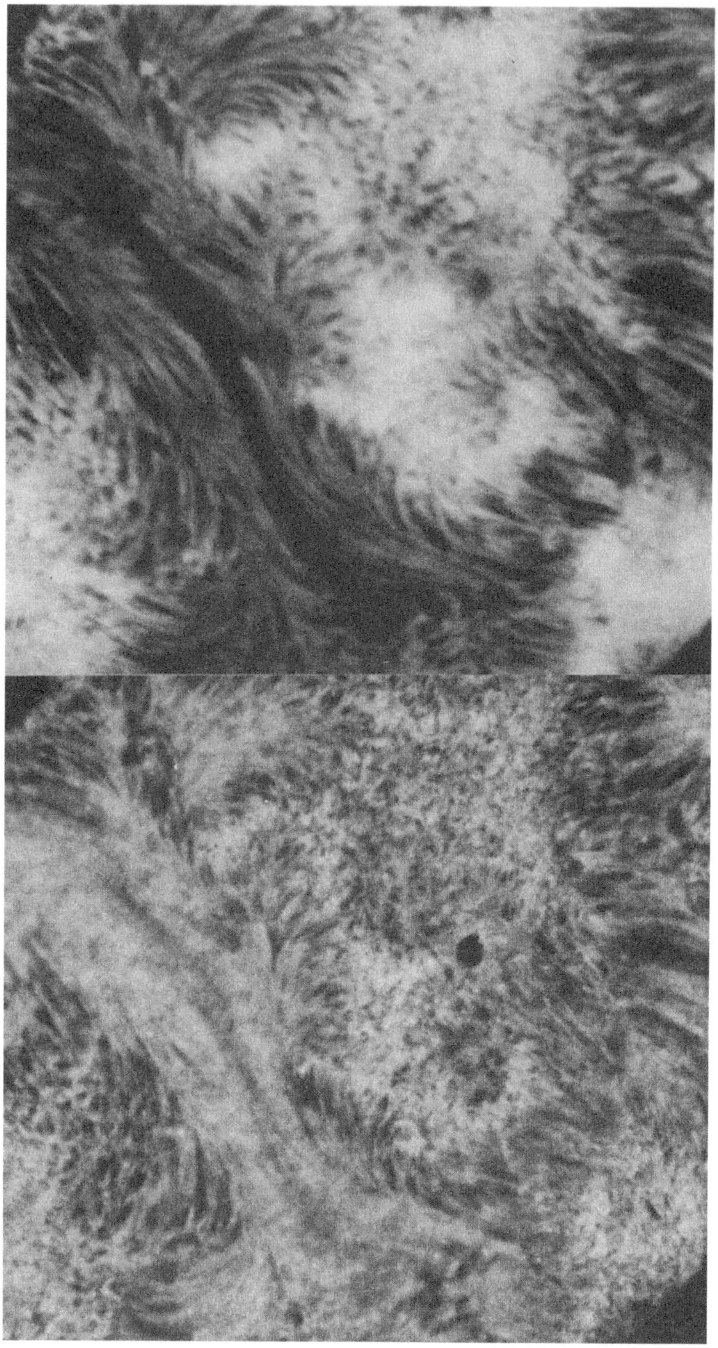

Fig. 7.6. *top*: Hα filtergram (line centre) with plage filament between Hα plages. Note fine structure of plage with fine dark mottles, and fibrils running along the filament; *bottom*: (Hα −0.5 Å) – filtergram: filament and fibrils replaced by filament channel (plage couloir). Note the fine mottles in the plage region. (Fraunhofer Institut, Anacapri Observatory.)

close (large) spots of opposite polarity. The filament is large and the channel is broad between weak plage fields of old decaying AR.

Martres, M. J., Michard, R., and Soru-Iscovici, I.: 1966, *Ann. Astrophys.* **29**, 249.

7.12. Chromospheric Plage

The (chromospheric) plage (originally 'plage faculaire') is an extended emission region observed in strong chromospheric lines such as Hα, Ca II H+K, H Lα, He II 304 Å as well as in lines of the transition region (Figures 7.1b,c). Above the plage – and coinciding with it in projection on the central parts of the solar disk – is the coronal condensation or enhancement emitting enhanced EUV, X-ray and radio emission. Its projected area is sometimes also called plage (EUV, X-ray, radio, coronal plage) (Figures 6.9, 7.1c,d). The chromospheric plage lies above the photospheric faculae.

These bright regions at photospheric, chromospheric and coronal level are regions of strong vertical magnetic field component. Estimated field strengths are up to 800 G. They are furthermore hotter and have (in the corona) a considerably higher density than the surrounding quiet regions.

They are the optical feature characteristic of AR's which exist from the very beginning of the AR (the emergence of first flux) until the widely scattered remnant magnetic fields merge with the background. The plage is brightest and densest in the young AR without visible fine structure. Aging plages (after maximum AR development, when no new flux emerges) as well as the old plages without any spots reveal a fine structure whose elements are usually referred to as *plage granules*. These have diameters $\approx 1''$ and separations $\approx 1\overset{''}{.}5$. Their lifetime is not yet known. Their exact relation to the elements of the photospheric facular fine structure has not yet been studied.

Bray, R. J. and Loughhead, R. E.: 1974, *The Solar Chromosphere*, Chapman and Hall, London, p. 255.
Zirin, H.: 1974 in R. G. Athay (ed.), 'Chromospheric Fine Structure', *IAU Symp.* **56**, 161.

7.13. Fibrils

Fibrils are one of the most conspicuous and characteristic Hα features in and around active regions. They are long, thin, dark streaks with a width of 725–2200 km and an average length of 11 000 km. The overall fibril pattern shows little change over a period of hours although the individual fibrils have a lifetime of only 10–20 min. (Figure 7.7).

In the central parts of AR's large numbers of fibrils are arranged in a pattern apparently connecting spots and plages of opposite polarities. Among them, *arch filaments* are a specially marked type existing only during the emergence of new flux. After several days, when flux has ceased to emerge, they are replaced by *field transition arches* between the spots; they are normal fibrils (Figures 7.3, 7.7).

Old, large, regular spots are surrounded by a radial fibril pattern called the (chromospheric) *superpenumbra* (Figure 7.7). The individual fibrils begin within the penumbra; the pattern extends to a distance of about a spot diameter. The large-scale pattern is remarkably stable for hours, but only a few individual fibrils remain identifiable as long as

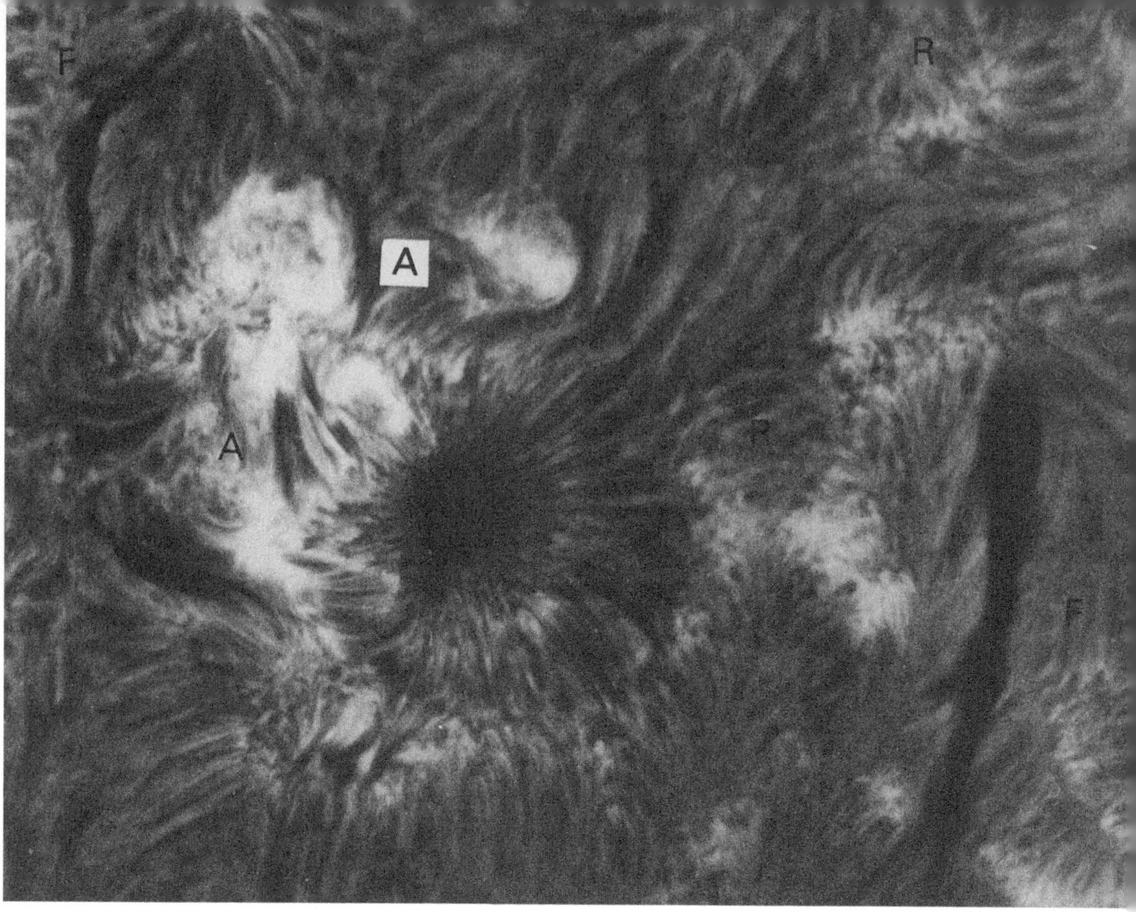

Fig. 7.7. Large regular spot with superpenumbra. Note also the field transition arches (*A*) and the other fibril patterns, the filaments (*F*) and the rosettes (*R*) with dark mottles. (Hα filtergramm, Fraunhofer Institut, Anacapri Observatory.)

20 min. In superpenumbral fibrils material is flowing inwards and downwards into the spot with $v \approx 20$ km s^{-1} (*inverse chromospheric Evershed effect*).

In the outer parts of an AR numerous fibrils running more or less radially or slightly curvilinear outwards form a pattern which is sometimes referred to as *solar vortex*. It defines a roughly elliptical area whose boundary moves continuously outwards with $v \approx 0.2$ km s^{-1} during the development of the active region (Figure 7.1b). This fibril pattern is visible also in high-resolution K photographs. In low-resolution K_3 spectroheliograms it appears as a very faint dark ellipse which has been called *circumfacule* by Deslandres.

Some authors argue that the fibrils follow the magnetic field in the chromosphere and therefore may be used in evaluating the configuration of the magnetic field in and around the active region. Immediately adjacent to filaments, fibrils run parallel to them indicating a magnetic field of the same orientation (Figure 7.6).

Giant fibrils or chains of fibrils forming very long, curved connections between large spots and/or plages are called *threads* They often resemble small active region filaments but, of course, do not follow a magnetic inversion line.

Bray, R. J. and Loughhead, R. E.: 1964, *Sunspots*, Chapman and Hall, London, p. 253.

Bray, R. J. and Loughhead, R. E.: 1974, *The Solar Chromosphere*, Chapman and Hall, London, p. 217.
Zirin, H.: 1974, in R. G. Athay (ed.), 'Chromospheric Fine Structure', *IAU Symp.* **56**, 161.

7.14. Moustache (Ellerman Bomb)

Moustaches (first detected by Ellerman and called bombs) appear as bright points (diameters from 3″ down to the resolution limit) on filtergrams taken in the wings of Hα and K. The Balmer series up to H_{10}, the H and K lines and a number of strong to weak metal lines have been observed in moustache spectra. Strong lines have a typical profile with absorption in the line centre and extended emission in the wings with a maximum contrast in Hα around 1.0 Å (0.75–2.0 Å) from the line centre. This unique type of profile (Figure 7.8) led the Russian observers to call the phenomenon YCOB, that is moustache. The line wings extend out to 5–10 Å and appear usually superimposed on thin bands of continuous emission extending through the entire visible spectrum; these bands are called *grains of continuous emission.* These grains exist also without moustaches and probably are the continuous spectrum of facular points or elements of the filigree.

Moustaches occur in active regions throughout their active phases, in particular in young regions in the area of AFS, that is, in emerging flux regions. They are found frequently close to small or large spots and at the feet of small, short-lived surge-like features (Figure 7.9). Rust claims that they coincide with isolated peaks of magnetic field

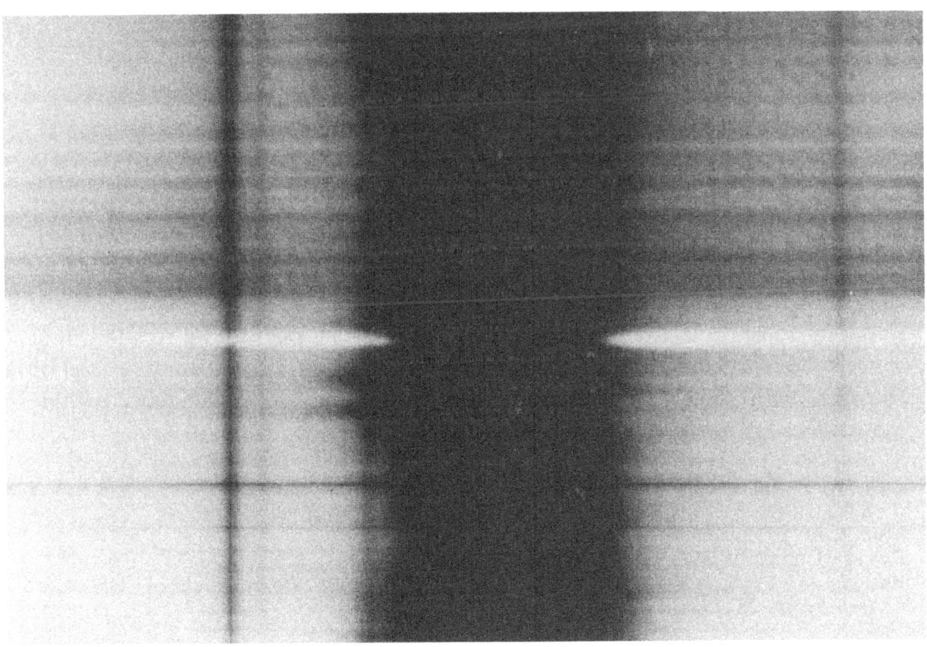

Fig. 7.8. Hα spectrum of a moustache with wide emission wings. (Fraunhofer Institut, Anacapri Observatory.)

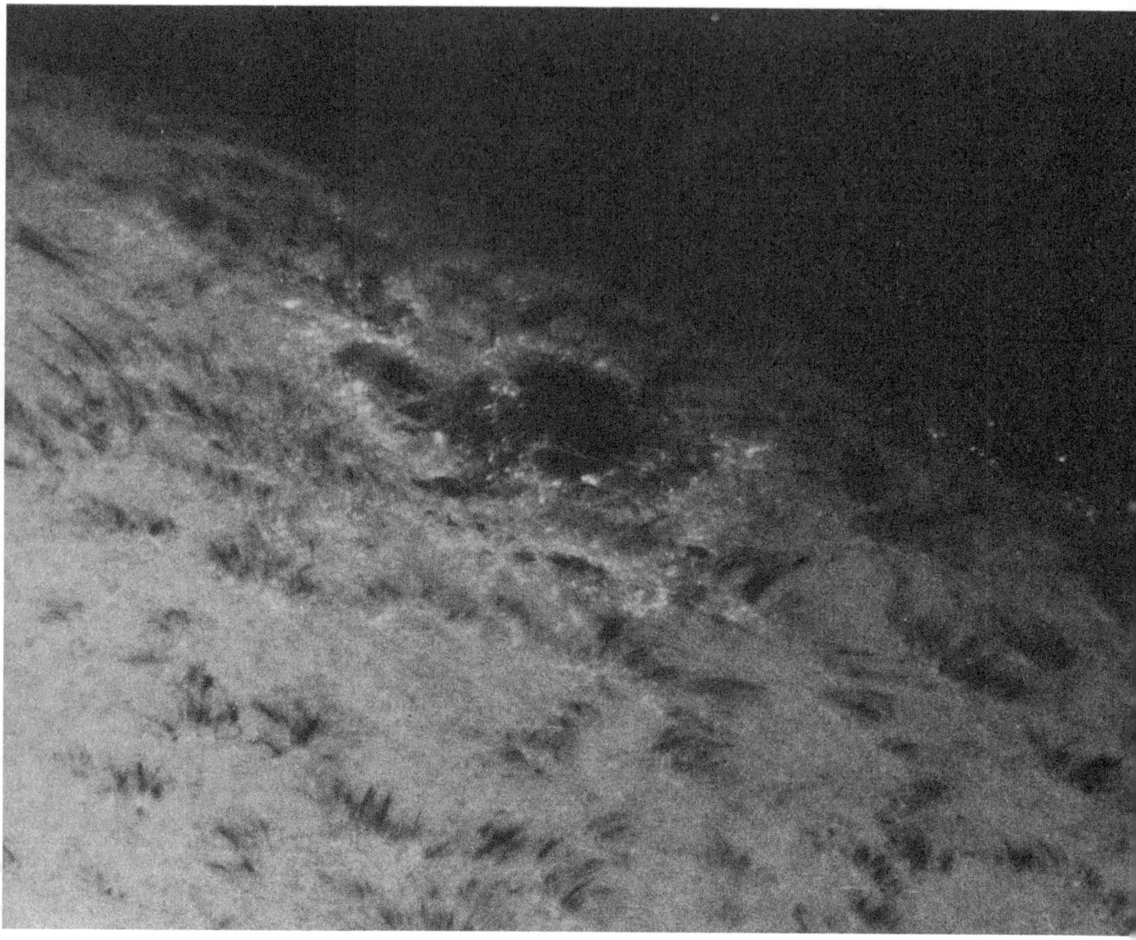

Fig. 7.9. Moustaches (bombs) around sunspots. Note also the 'bushes' outlining chromospheric network cells. (Photograph: Big Bear Solar Observatory.)

('satellite sunspots'). Their lifetimes range from several minutes up to a few hours (typical value ≈ 20 min); they usually brighten within 2–3 min, then keep their level of intensity (possibly with some fluctuation) and finally disappear again suddenly within 2–3 min. Moustaches are not generally related to flares.

Bray, R. J. and Loughhead, R. E.: 1974, *The Solar Chromosphere*, Chapman and Hall, London, p. 227.
Bruzek, A.: 1972, *Solar Phys.* **26**, 94.
Roy, J.-R. and Leparskas, H.: 1973, *Solar Phys.* **30**, 449.
Rust, D. M.: 1968, in K. O. Kiepenheuer (ed.), 'Structure and Development of Solar Active Regions', *IAU Symp.* **35**, 77.

7.15. Magnetic Classification of Active Regions

Detailed magnetic classifications have been introduced which characterize the configuration of the magnetic field in active regions taking into account not only the distribution of the spot fields but also of the surrounding weaker fields of faculae (plages). Essentially, they characterize the degree of complexity which is related statistically to the flare productivity of the active region. For details see the publications given below.

Martres, M. J., Michard, R., and Soru-Iscovici, I.: 1966, *Ann. Astrophys.* **29**, 245.
Smith, S. F. and Howard, R.: 1968, in K. O. Kiepenheuer (ed.), 'Structure and Development of Solar Active Regions', *IAU Symp.* **35**, 33.

7.16. Mt. Wilson Spot Classification

This is a classification according to the configuration of the magnetic field of the spots in the active region. It comprises three main classes:

α = unipolar, β = bipolar, γ = complex.

α and β are subdivided into a number of subclasses:
α: the spot (group) is in the centre of the AR (the *K* plage),
αp and αf: the spot is in the preceding and the following part of the AR, respectively,
β: the spotgroup is virtually symmetric,
βp and βf: the *p*- and *f*-spots dominate, respectively,
$\beta\gamma$: virtually bipolar but no clear dividing line between north and south polarities.
γ: the polarities are so irregularly distributed as to prevent classification as bipolar groups.
$\beta\gamma$ and γ groups are the most flare-active spot classes.

As additional characteristics 'local' magnetic classifications have been introduced which denote the existence of close umbrae of opposite magnetic polarities in a part of the spot group:

The *δ-configuration* – introduced by Künzel – refers to the existence of umbrae of opposite polarities within one penumbra in β, $\beta\gamma$ and γ groups.

Configuration A – proposed by Avignon *et al.* – comprises two spots or two rows of spots of opposite polarities with the opposite umbrae (sometimes in the same penumbra) at a distance of less than 1 heliographic degree. The spots of both polarities cover approximately the same area. This configuration is called A' if these areas are different and/or a Hα filament exists on the inversion line. The configurations δ and A are of special interest because they are most favourable for the occurrence of powerful flares.

Another class is *configuration B* which consists of an old, unipolar spot completely surrounded by a plage region of opposite polarity.

In order to avoid confusion it should be noted that *Solar Geophysical Data* uses latin capitals throughout for the Mt. Wilson classification: *A* for unipolar, *B* for bipolar, *C* for complex spot groups and *D* for the δ-configuration; accordingly *AP, AF, BP* etc.

Avignon, Y., Martres, M.-J., and Pick, M.: 1964, *Ann. Astrophys.* **27**, 23.
de Jager, C.: 1959, in S. Flügge (ed.), *The Solar System*, (Encyclopedia of Physics, Vol. LII), p. 166.
Künzel, H.: 1960, *Astron. Nachr.* **285**, 271.

7.17. Solar Activity Indices

The general, global solar activity refers to the abundance and size of active regions on the visible solar disk and is characterized by various *activity indices.* These are:

(a) *Wolf Number* or *Zürich Sunspot Relative Number* defined as $R = k(f + 10\,g)$, f = total number of spots on the visible hemisphere, g = number of spot groups, k reduces the observed values to the standard Zürich number.

A *Central Relative Number* R_z is defined in the same way for the central region of the solar disk with $r = \frac{1}{2} r_0$.

Monthly and yearly means, and monthly means smoothed over 13 months are used in correlation studies.

(b) the total area covered by sunspots on the visible hemisphere (Greenwich);

(c) *K-plage index* (McMath-Hulbert Observatory) is derived from plage areas and brightness.

Additionally, the global radio and X-ray flux are used as indices of solar activity.

Solar activity varies with a period of approximately 11 yr (see Solar Cycle).

7.18. Solar Activity Data

Data on solar activity are published regularly by a number of institutes. The most important publications are:

(1) *Solar Geophysical Data*, NOAA/ESSA, Boulder. Published since 1945 by different institutions and with varying content, this is at present the most comprehensive and fastest publication of solar activity and related interplanetary and terrestrial data. It consists now of two parts: I. Prompt Reports (with one and two months delay), II. Comprehensive Reports (6 months delay) which give extensive data on active regions (all types of optical observations, X-ray and cm heliograms, magnetograms, flares), global radio intensities and radiobursts, X-ray intensity, cosmic rays, solar wind and protons, interplanetary magnetic and electric fields, ionospheric and geomagnetic disturbances.

(2) *Solnechnye Dannye* (Solar Data), Academy of Science, Leningrad, U.S.S.R., since 1956 published with a delay of about 6 months. It includes daily maps of the Sun showing spots (with field strengths), plages, filaments and tables of spot groups, flares, ejections, radio intensities, bursts.

(3) *Monthly Bulletin on Solar Phenomena* + Photographic Journal of the Sun, Osservatorio di Roma, since 1958 (1958-1964: Circolare Fenomeni Solari) presents data on spots (position, area, magnetic field strengths).

(4) *Quarterly Bulletin on Solar Activity*, IAU – Eidgenössische Sternwarte, Zürich, since 1917 (until 1938: Bulletin for Character Figures of Solar Phenomena); delay ≈ 18 months. It provides at present: Wolf numbers and spot areas, synoptic magnetic field charts, lists of flares, coronal line intensities, global radio intensities, distinctive radio events (bursts), radio activity chart.

(5) *Photoheliographic Results*, Royal Greenwich Observatory 1874–1976 presents the longest and most complete series of spot data (positions and areas of spot groups and of their main spots).

(6) *Heliographische Karten* der Photosphäre, Eidgenössische Sternwarte Zürich, since 1897; synoptic charts of active regions (spot groups and photospheric faculae) and tables of spot development.

(7) *Cartes synoptiques* de la Chromosphère Solaire et Catalogue des Filaments et Centres d'Activité, Observatoire de Paris-Meudon, since 1919.

8. SPOTS AND FACULAE

A. BRUZEK

8.1. Photospheric Faculae

These are extended regions appearing bright in photospheric lines (of neutral species) and in the continuum. There is a smooth transition from the bright, dense faculae surrounding sunspots in young and mature active regions, through faint, scattered faculae of spotless remnants of active regions to the weak photospheric network. The faculae are cospatial with bright plages observed in chromospheric lines (H Balmer lines, Ca II lines, He II 304 Å) and transition region lines, and with regions of enhanced magnetic fields.

The structure can be traced through all layers of the solar atmosphere though the elements tend to coarsen with increasing height. They are subarcsecond in the photosphere, several arcseconds in the chromosphere (Ca II K line core) and ≈ 15 000 km in the transition region (Mg X line). This suggests a divergence of the magnetic field patches associated with the faculae-plage elements.

At visible-continuum wavelengths faculae are best seen near the solar limb (Figure 8.1). There, medium-resolution observations show apparent *facular granules* with

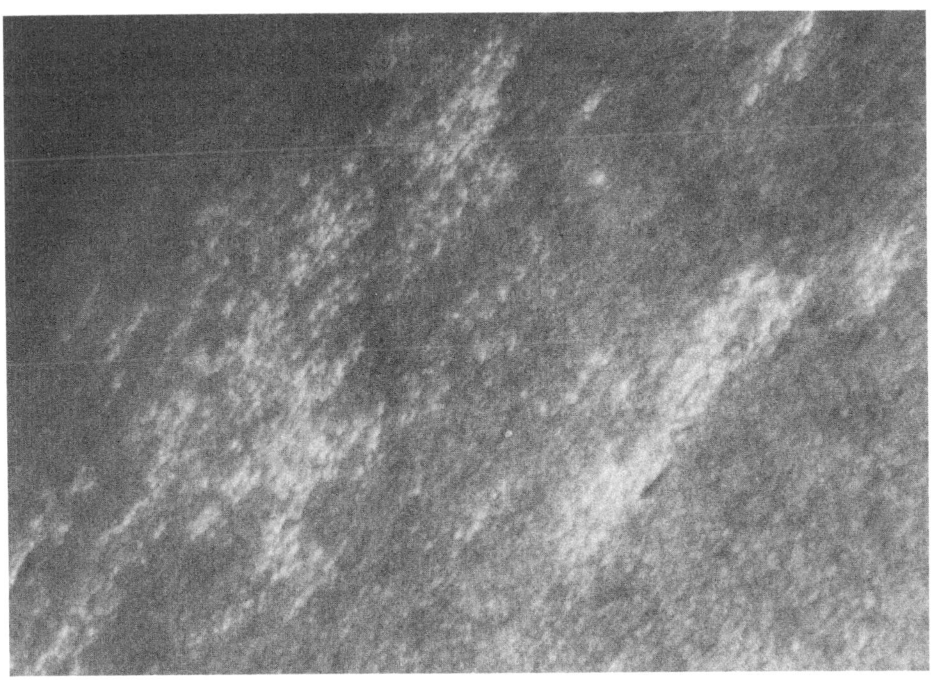

Fig. 8.1. Photospheric faculae resolved into facular granules. (Photograph: Observatoire du Pic du Midi. Courtesy R. Muller, May 31, 1973.)

Bruzek and Durrant (eds.), Illustrated Glossary for Solar and Solar-Terrestrial Physics. 71–79.

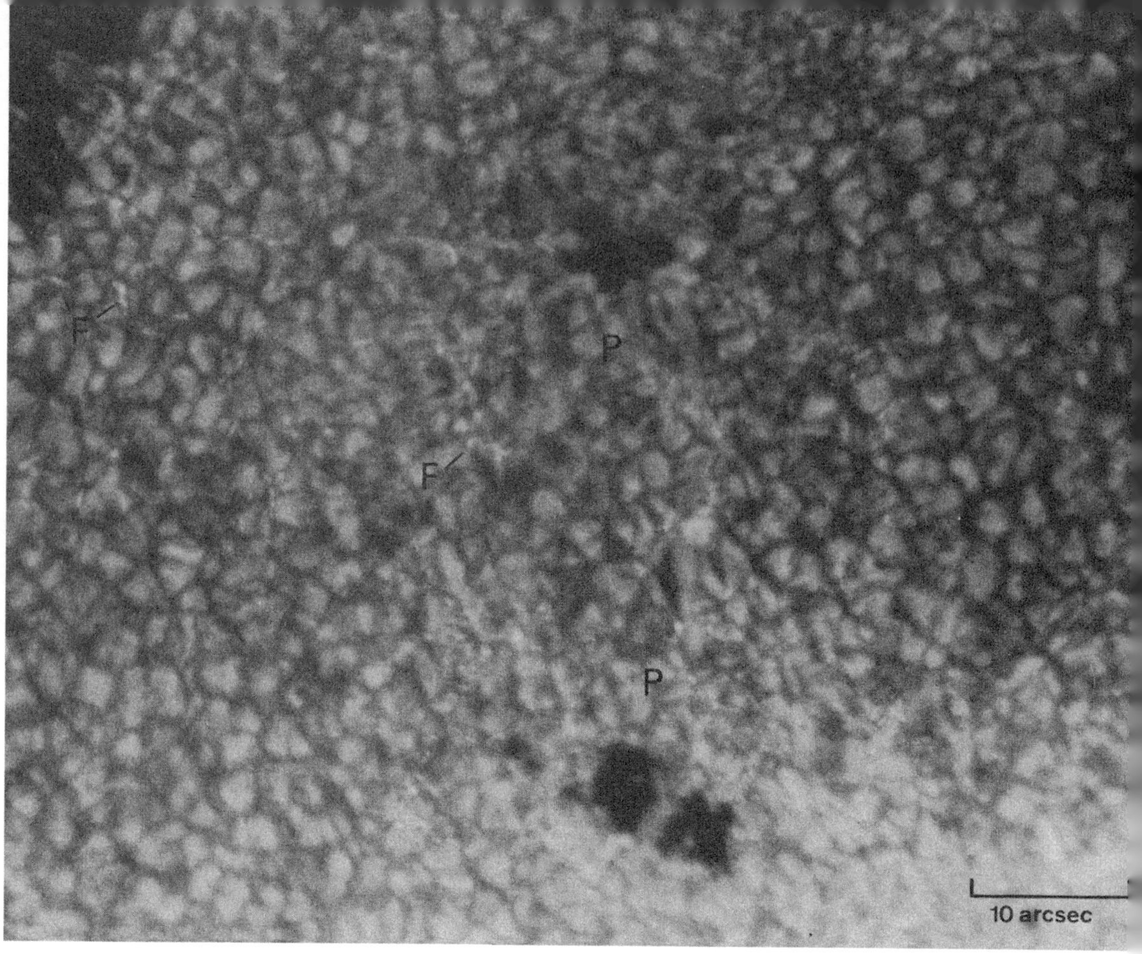

Fig. 8.2. Filigrees (*F*) and pores (*P*) in the granular pattern. Photographed in the centre of the disk April 28, 1973 at λ3934 ± 30 Å. (Sacramento Peak Observatory, Courtesy J. P. Mehltretter.)

diameters of 1–2″. They probably correspond to clusters of unresolved facular points. The facular life-time of two hours quoted in the 1960's would refer to such clusters.

The unresolved structure of faculae renders uncorrected values of their centre-to-limb contrast and models based on them obsolete.

Bray, R. J. and Loughhead, R. E.: 1961, *Australian J. Phys.* **14**, 14.
Mehltretter, J. P.: 1974, *Solar Phys.* **38**, 43.

8.2. Facular Point

In regions far from spots facular structure in the wings of the Ca II K line appears almost exclusively to be made up of bright facular points sitting in the intergranular lanes – singly or associated in chains. Close to spots it consists mainly of *crinkles* with a ring-like formation; existing observations do not show whether they are in fact associations of 'points'. In the far wing of Hα (Hα + 2 Å) and nearby continuum a similar crinkle pattern – the *filigree* – is observed which is tentatively identified with the Ca II pattern. (See Figure 8.2.)

The width of the points and crinkles is at the limit of resolution ($\approx 0.25''$), and crinkles extend up to $\approx 2.5''$ in length. The contrast in the K line wing ranges between 0.26 and 0.98 (corrected values). Most points seem to brighten and fade within 2–3 min but to be present for intervals of 5–15 min. Some associations of points have lifetimes $\approx \frac{1}{2}$ hr.

The general location of the filigree is stable, but the individual crinkles are jostled around over a distance of the order of a granule diameter during 40 min with a velocity ≈ 1.5 km s^{-1}. This suggests that the granules have enough energy to dislocate crinkles and hence the local magnetic field. It is an indication that crinkles are associated with magnetic fields weaker than in pores.

The number density of facular points varies from 160 points per $10''$ square in active regions to 6 per $10''$ square outside active regions.

If it is assumed that all magnetic flux is concentrated in these structures they would possess an average flux of 4.4×10^{17} Mx and a mean field strength ≈ 2500 G.

Dunn, R. B. and Zirker, J. B.: 1973, *Solar Phys.* **33**, 281.
Mehltretter, J. P.: 1974, *Solar Phys.* **38**, 43.

8.3. Polar Faculae

Polar faculae appear as isolated facular structures with diameters $\approx 3''$ in the polar belts around latitudes 67° north and south. They exist only during a 1–2 yr period preceding the solar minimum and probably are associated with polar coronal rays. Their mean lifetime is 15 min. They are visible also in strong chromospheric lines (Hα, K of Ca II).

Waldmeier, M.: 1955, *Z. Astrophys.* **38**, 37.

8.4. Magnetic Knots

Magnetic knots are aggregates of strong magnetic field (strength 1400 or even 2000 G). They are coincident with small (≈ 1000 km) regions where certain lines of neutral metals appear shallower and broader producing a *line gap*. The gap is caused by strong Zeeman splitting and significantly higher temperatures in the line forming regions.

Magnetic knots appear in large numbers in the vicinity of sunspots but also occur in the network far from spots. Around spots their number density is 8–10 knots per 100 granules up to distances of 8×10^4 km from the spot. The magnetic flux in a typical knot is 8×10^{18} Mx. The total flux in knots surrounding a spot is 14×10^{21} Mx for knots of opposite polarity to the spot and 7×10^{21} Mx for knots of the same polarity. The net flux equals the total flux of a typical spot but is of the opposite polarity suggesting that nearly all the magnetic flux from sunspots returns to the photosphere through the knots. Knots are associated with strong downward motions in the photosphere and live for about 1 hr. The majority of knots lie above intergranular regions of the continuum and below Ca II K_{232} emission (plages and network).

Their relationship to moving magnetic features (MMF), magnetic elements and invisible sunspots is not clear.

Beckers, J. M. and Schröter, E. H.: 1968, *Solar Phys.* **4**, 142.
Sheeley, N. R.: 1967, *Solar Phys.* 1, 171.

8.5. Moving Magnetic Features (MMF)

There are two types of small ($< 2''$) moving magnetic elements (or clumps of elements) which are referred to as MMF: one type is associated with *magnetic flux outflow* (MFO) from spots, the other with *magnetic flux inflow* (MFI) into spots or more generally towards a concentration of flux. MMF's are associated with sunspots of all sizes and degrees of complexity (Figure 7.2).

MFO's are observed mainly with decaying sunspots which are surrounded by a *moat*, Figure 7.2, an annular zone devoid of stationary fields extending 10 000–20 000 km from the spot's edge. The MFO's move radially away from the spot towards the nearby magnetic network with constant velocities from a few tenths to 2.0 km s^{-1}. They are visible as bright points in CN and K_{2v} spectroheliograms, but appear only in very faint emission in the core of Hα. Individual MFO's may have either north or south polarity, but altogether they show a net magnetic flux of the same sign as the parent spot. The average flux of MFO's was found to be about 10^{19} Mx; 6×10^{17}–8×10^{19} Mx is the range of individual values.

The moat thus has the physical aspect of a single, outwelling velocity cell – similar to a supergranulation cell – centred on the sunspot. MFO's appear to be the result of the flux rope fragmentation responsible for spot decay.

Magnetic flux inflow is a phenomenon manifested by MMF's that stream towards a sunspot along paths converging upon it. In contrast to MFO's many, but not all, of the MMF's involved in MFI are pores easily observed in integrated light, some of them moving directly into the umbra of a spot at speeds of 0.25–1.0 km s^{-1}.

MFI's are associated with emerging flux regions and with growing spots. The magnetic elements moving towards a spot all have the same magnetic polarity as that of the spot.

Frazier, E. N.: 1972, *Solar Phys.* **26**, 130.
Harvey, K. and Harvey, J.: 1973, *Solar Phys.* **28**, 61.
Vrabec, D.: 1974 in R. G. Athay (ed.), 'Chromospheric Fine Structure', *IAU Symp.* **56**, 201.

8.6. Pore

Pores are small sunspots without penumbrae. Their diameters are in the range $1''$–$5''$; their brightness is about 50% photospheric brightness. The lifetime of pores is of the order of 1 day; they remain virtually unchanged for many hours. They are formed within about 45 min but little is known about their birth. Only a few pores develop into full-size spots. The magnetic field associated with pores is > 1500 G. See Figure 8.2.

Pores are abundant in sunspot groups but are also found far removed from them. Little information is available about the exact distribution of pores over the solar disk.

Micropores are dark spots much smaller than pores which are visible only on pictures having the very best resolution. Therefore little is known about their characteristics. They are ringed by filigree and many seem to be produced by the absence of faculae in one point of intergranular space. Larger micropores, however, are decidely darker than intergranular space and have lifetimes of 10–20 min. These may be identical with magnetic knots. There are no magnetic field measurements in micropores.

Bray, R. J. and Loughhead, R. E.: 1964, *Sunspots*, Chapman and Hall, London, p. 69.
Dunn, R. B. and Zirker, J. B.: 1973, *Solar Phys.* **33**, 281.
Mehltretter, J. P.: 1974, *Solar Phys.* **38**, 43.

8.7. Sunspot

Sunspots appear as photospheric regions with decreased temperature, radiation and gas pressure and are concentrations of magnetic flux with field strengths of 2000–4000 G. They consist of one or more dark cores ('umbra') surrounded by a less dark penumbra. Spots have diameters $\gtrsim 10''$ up to $\approx 1'$. Individual spots grow at a rate of up to ≈ 100 'millionths' (of the area of the solar hemisphere) per day; the decay rate of long-lived, regular spots is a constant 6 millionths per day. Spots form groups which contain up to several tens of individual spots of either magnetic polarity in a bipolar or multipolar arrangement. The magnetic flux of a large spot is $\approx 10^{21}$ Mx, that of large spot groups $\approx 10^{22}$ Mx (Figure 8.3).

The gas pressure in spots is reduced considerably because gas plus magnetic pressures

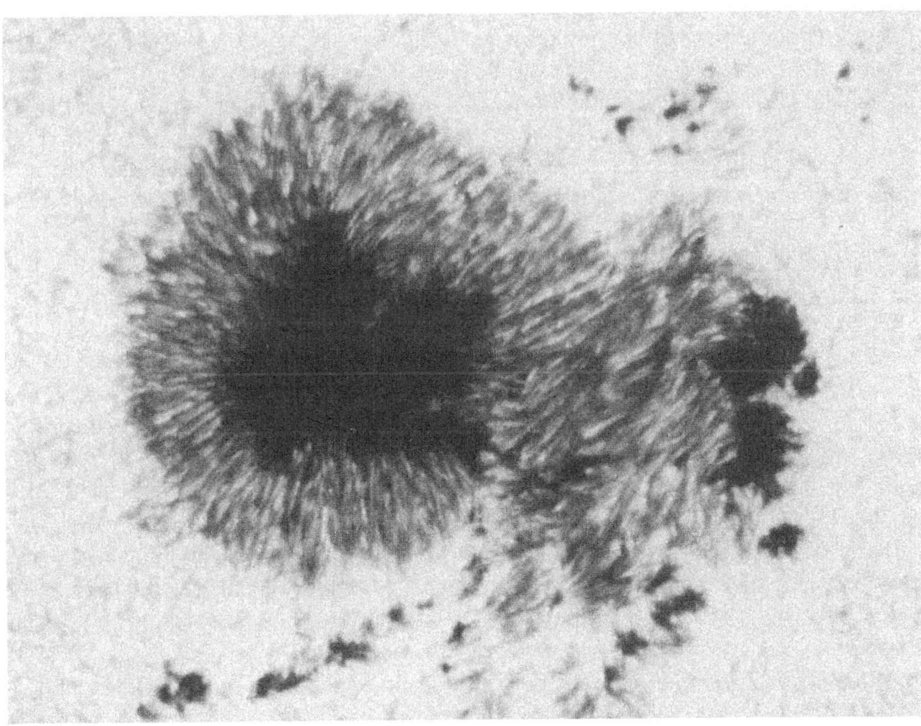

Fig. 8.3. **Large sunspot with fine structure** of penumbra (filaments, grains). (Observatoire du Pic du Midi, July 5, 1970, λ5280 ± 50 Å; Courtesy R. Muller.)

balance the gas pressure in the surrounding photosphere. The reduced temperature of the spot is due also to the presence of the strong magnetic field; there is, however, no general agreement on the effective mechanism. According to Biermann's original theory (which has been worked out by a number of authors) convective transfer of energy below the photospheric regions is reduced ('inhibited') by a strong magnetic field; therefore a deficit in radiative flux occurs in the visible spot amounting to 4×10^{29} erg s^{-1} in a large spot group. Then there arises the *missing flux problem*, i.e. the question of what happens to the flux missing in the spot; where does it reappear? A possible answer is that it is redistributed in a very extended region around the spot. Parker, on the other hand, has suggested recently that a sunspot is a region of considerably enhanced mechanical energy flow because the magnetic field converts at least three quarters of the heat flux into hydromagnetic waves which propagate rapidly along the field through the spot without dissipation.

Bray, R. J. and Loughhead, R. E.: 1964, *Sunspots*, Chapman and Hall, London.
Parker, E. N.: 1974, *Solar Phys.* **36**, 249.
Wilson, P. R.: 1972, *Solar Phys.* **27**, 354, 363.

8.8. Umbra

The umbra is the dark core in a sunspot; it covers 0.17 (average value) of the total area of the spot. The intensity of the umbra is reduced to 5–15% of the photospheric intensity in visible light, dependent on the wavelength (increasing from violet to red light) but independent of the size of the umbra. The effective temperature of the umbra is $T_e = 3700$ K; its spectral type is K3–K5.

A remarkable degree of activity (fine structure, motions) has been detected in large umbrae in recent years; see: umbral dots, umbral oscillations, umbral flashes.

Ekmann, G. and Maltby. P.: 1974, *Solar Phys.* **35**, 317.
Mattig, W.: 1971, *Solar Phys.* **18**, 434.
Wöhl, H., Wittmann, A., and Schröter, E. H.: 1970, *Solar Phys.* **13**, 104.

8.9. Umbral Dots

Umbral dots are very small, bright continuum points in sunspot umbrae with estimated depths of about 100 km and with upward motion. Their true diameters are 150–200 km, their lifetime is 1500 s, and they have an estimated temperature of 6300 K. The magnetic field seems to be sharply decreased or even reversed in umbral dots. Joule heating associated with such magnetic inhomogeneities may be responsible for the dots. Some 20 umbral dots may be found at any one time in a medium-sized umbra. There is no indication of a relationship between the position of umbral dots and that of penumbral filaments. *Umbral granulation* is the poor-resolution appearance of umbral dots.

Beckers, J. M. and Schröter, E. H.: 1968, *Solar Phys.* **4**, 303.
Kneer, F.: 1973, *Solar Phys.* **28**, 361.
Krat, V. A., Karpinsky, V. N., and Pravdjuk, L. M.: 1972, *Solar Phys.* **26**, 305.

8.10. Umbral Flashes

These are rapidly changing, bright inhomogeneities visible in the H and K lines and the infrared triplet of ionized calcium. They have a typical lifetime of 50 s with a fast increase and slow decrease in brightness; they have a tendency to repeat every 145 s. They show a Doppler shift of ≈ 6 km s^{-1} and move towards the penumbra with a velocity ≈ 40 km s^{-1}. The exciting agent seems to have a vertical velocity ≈ 90 km s^{-1}. The diameters of the flashes are ≈ 2000 km and magnetic fields of 2000 G are observed (in the K line). However, in one case, the Zeeman splitting of the K line indicated a field strength of 5500 G. It is not known whether the umbral flashes are related to the umbral dots. It has been suggested that the flashes are produced by magneto-acoustic waves arising from lower levels in the umbra.

Umbral flashes are sometimes seen also in Hα, but appear more diffuse than in H and K. Moreover, while K line flashes result from actual increases in line emission, there is evidence that the Hα flashes are primarily due to Doppler shifts of the Hα absorption line profile produced by the chromospheric umbral oscillations.

Beckers, J. M. and Tallant, P. E.: 1969, *Solar Phys.* **7**, 351.
Moore, R. L. and Tang, F.: 1975, *Solar Phys.* **41**, 81.

8.11. Umbral Oscillations

In the umbral photosphere vertical velocity oscillations with periods of 165 ± 20 s and typical amplitudes of 0.2 km s^{-1} are observed. In the umbral (Hα) chromosphere the oscillation period seems to be somewhat shorter whereas the velocity amplitude ranges between one and six km s^{-1}. The horizontal size of an oscillating element is a few hundred up to 2000 km. The oscillations have a duration of 8 cycles or more. They can be explained as the result of either a standing wave in the overstable layer or a standing wave in the resonance cavity between this layer and the chromosphere-corona transition region.

Beckers, J. M. and Schultz, R. B.: 1972, *Solar Phys.* **27**, 61.
Giovanelli, R. G.: 1972, *Solar Phys.* **27**, 71.

8.12. Light Bridges

Light bridges are bright tongues or streaks penetrating or crossing spot umbrae. Their brightness ranges from normal photospheric to facular values. Sometimes they appear clearly resolved into one or two rows of bright grains or granules which may be facular material. A relationship to faculae is suggested also by the observation that light bridges are most prominent in spots near the solar limb.

Light bridges develop at a slow rate and have lifetimes of several days. In some cases they persist throughout the entire lifetime of the spot. The appearance of a light bridge during the later part of the life of a spot is frequently a sign of impending division or final dissolution.

Bray, R. J. and Loughhead, R. E.: 1964, *Sunspots*, Chapman and Hall, London.
Vazquez, M.: 1973, *Solar Phys.* **31**, 377.

8.13. Penumbra

The penumbra is the outer, brighter part of a sunspot which – at lower resolution – appears to be made up of bright and dark *penumbral filaments.* Only high resolution photographs reveal that the penumbra consists of *bright grains* which generally are aligned forming narrow, bright filaments on a darker background. The penumbral bright grains are usually elongated features with a length between 0.5″ to 2.0″ and a width $\leqslant 0.5''$. The bright filaments have a distance 0.5″ to 1.0″ from each other and run radially outwards in regular spots. (See Figure 8.3.)

The grains form all over the penumbra and move towards the umbra of the spot with a horizontal velocity reaching a maximum at the umbral border ($v_{max} \approx 0.5$ km s^{-1}). Their lifetime ranges between 40 min and $\geqslant 3$ hr and depends on the place of origin. The average brightness of the grains at λ5280 Å is 0.95, referred to the photospheric brightness, while that of the dark background is 0.6. The corresponding temperatures at unit optical depth are 6310 K and 5715 K respectively in the outer parts of the penumbra.

Penumbral grains cover 43% of the penumbra. The mean brightness of the penumbra (averaged over bright and dark elements or measured in the unresolved penumbra) is 0.64 at λ3870 Å, 0.725 at λ5100 Å and increasing to 0.936 at λ3.8 μ (see also Evershed effect).

Maltby, P.: 1972, *Solar Phys.* **26**, 76.
Muller, R.: 1973, *Solar Phys.* **29**, 55; **32**, 409.
Wöhl, H., Wittmann, A., and Schröter, E. H.: 1970, *Solar Phys.* **13**, 104.

8.14. Penumbral Waves

In very regular spots running penumbral waves are seen developing just inside the umbral border and propagating outwards with a velocity ≈ 20 km s^{-1}. They are visible in Hα both as intensity and velocity features; the velocity amplitudes are $\approx \pm 1$ km s^{-1}, observed periods range between 210 and 270 s. The penumbral waves and the umbral flashes have no regular phase relation and therefore are probably physically independent. The penumbral waves have been interpreted alternatively as sound waves, Alfvén waves guided by the spot magnetic field, and as magneto-acoustic waves.

Also from the umbra border diffuse, cloud-like '*dark puffs*' are observed to emerge with the same period and the same rate as the penumbral waves. They appear to be superposed on the penumbral waves which can be seen only faintly below them. Likewise the filamentary structure of the penumbra appears to be obscured or absent in the dark puffs. The dark puffs are therefore interpreted as absorbing clouds which are higher in the chromosphere than the penumbral waves.

Giovanelli, R. G.: 1972, *Solar Phys.* **27**, 71.
Moore, R. L. and Tang, F.: 1975, *Solar Phys.* **41**, 81.

Nye, A. H. and Thomas, J. H.: 1974, *Solar Phys.* **38**, 399.
Zirin, H. and Stein, A.: 1972, *Astrophys. J.* **178**, L85.

8.15. Evershed Effect

The Evershed effect is a Doppler shift observed in the penumbral spectrum of spots close to the solar limb. It is due to radial mass motions occurring nearly parallel to the solar surface. They are directed outwards in the photosphere as observed in weak lines (Evershed proper) and towards the umbra in the chromosphere observed in strong lines (inverse Evershed effect, see fibrils).

If bright and dark penumbral elements are not resolved an average photospheric Evershed velocity $\approx 2 \text{ km s}^{-1}$ is measured. High resolution spectra show, however, that outflow occurs in dark regions only, with $v \approx 6 \text{ km s}^{-1}$ while the bright grains show a slow inflow.

Abdusamatov, H. I. and Krat, V. A.: 1970, *Solar Phys.* **14**, 132.

8.16. Wilson Effect

In sunspots close to the solar limb the apparent width of the penumbra on the side remote from the limb is smaller than that on the limb side; the ratio of the both widths decreases with decreasing distance from the limb. This effect has been explained by spot models in which $\tau \approx 1$ in the umbra occurs ≈ 700 km below the level of $\tau \approx 1$ in the surrounding photosphere. The larger depth of the visible level in the umbra may be due either to reduced opacity or be a real geometrical depression.

Bray, R. J. and Loughhead, R. E.: 1964, *Sunspots*, Chapman and Hall, London, p. 93ff.
Mattig, W.: 1969, *Solar Phys.* **8**, 291.
Wilson, P. R.: 1968, *Solar Phys.* **5**, 338.
Wilson, P. R. and Cannon, C. J.: 1968, *Solar Phys.* **4** 3.

9. FLARES AND ASSOCIATED PHENOMENA

H. W. DODSON-PRINCE and A. BRUZEK

9.1. Solar Flare

The solar flare is the response of the solar atmosphere (mainly chromosphere and corona) to a sudden, transient release of energy (probably of magnetic origin) which leads primarily to a localized, temporary heating (= thermal flare) and to an acceleration of electrons, protons and heavier ions (= particle or high energy flare). The temperatures attained in the chromosphere are $\approx 10^4$ K (chromospheric or low temperature flare), in the corona $\approx 10^7$ K (high temperature flare). The energies of observed particles range from 20 keV up to $\approx$ 1 GeV. The total energy released in the largest events is $\approx 10^{32}$ erg.

The flare produces transient electromagnetic radiation over a very wide range of wavelengths extending from hard X-rays ($\lambda \approx 10^{-9}$ cm) – in very rare cases from gamma rays ($\lambda \approx 2 \times 10^{-11}$ cm) – to km radio waves (10^6 cm). The radiation is predominantly thermal in nature; only at very short wavelengths (hard X-rays, $\lambda < 1$ Å) and at very long wavelengths (radiowaves) do non-thermal, impulsive, short-lived bursts of emission occur, produced by energetic particles (non-thermal bremsstrahlung, synchrotron radiation, etc.) or by shock waves.

The different kinds of radiation originate in different regions of the solar atmosphere and are accordingly referred to as the Hα flare, X-ray flare, radio flare etc.

Flares are closely associated with active regions: the majority of them occur in young or mature active regions; large flares prefer regions with large spots, complex magnetic field configuration and high field gradients. However, a number of flares (7% of major flares) appear between active regions or in old spotless or nearly spotless regions with remnant plages. Many of these 'spotless' flares are associated with the disappearance of a quiescent filament.

Smith, H. J. and Smith, E. V. P.: 1963, *Solar Flares*, Macmillan, New York.
Švestka, Z.: 1976, *Solar Flares*, D. Reidel Publ. Co., Dordrecht.
Zirin, H.: 1974, *Vistas Astron.* **16**, 1.

9.2. Thermal Flare

The thermal flare has been best observed and studied in the Hα line of hydrogen and, in recent years and to a lesser degree, also in soft X-rays. The Hα flare and the X-ray flare are products of the low and high temperature flare, respectively. They have quite a number of characteristics in common so that, in many respects, the *Hα flare* stands for the thermal or optical flare in general.

The duration of the *optical flare* ranges from several minutes to a few hours. For almost all flares, the rise to maximum intensity in Hα radiation is more rapid than the decline. The time of rise to maximum may be only a few minutes, or as long as an hour for *'slow' flares.*

Bruzek and Durrant (eds.), Illustrated Glossary for Solar and Solar-Terrestrial Physics. 81–96.

For the majority of flares and subflares the maximum intensity of Hα is less than that of the local continuous spectrum, but for the brightest events the Hα radiation may be approximately twice as bright as the intensity of the continuum. The Balmer (and other strong chromospheric) lines are very wide in some flares, up to 10 Å (Figure 9.1). The Hα line profile is generally asymmetrical i.e. the one wing appears broadened and/or intensified compared to the other. *Red-asymmetry*, i.e. predominance of emission $>$ Hα (centre) prevails, in particular in the flash phase. No satisfactory interpretation of the line asymmetry has been given yet.

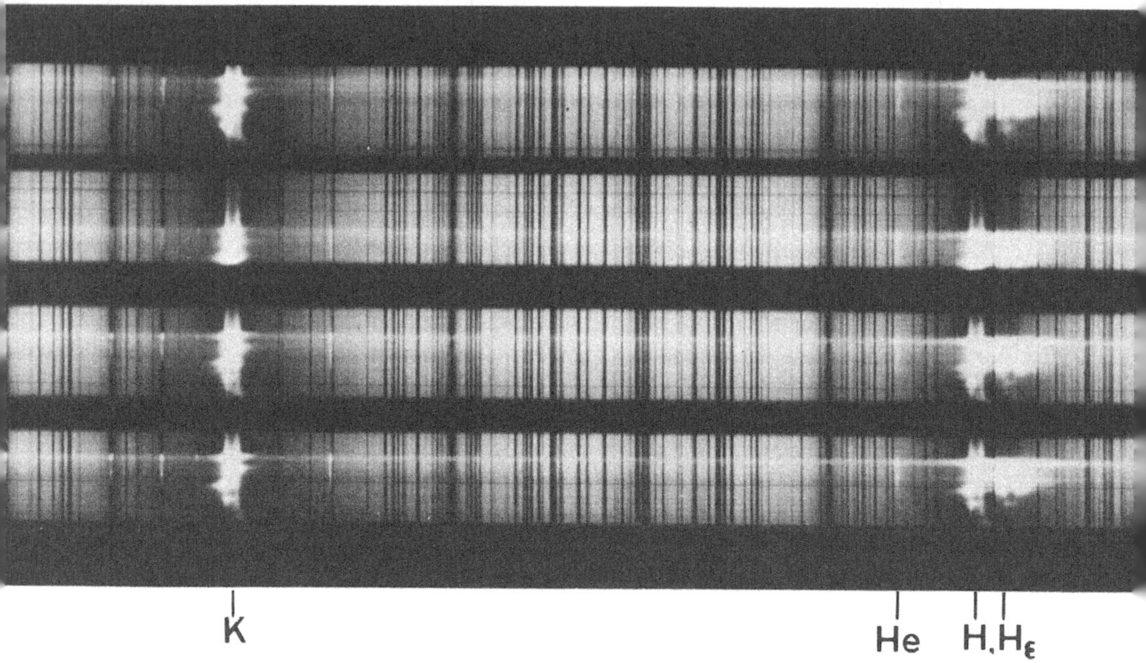

Fig. 9.1. Flare spectra in the region of the K and H Ca II lines. (March 12, 1969, Observatorio de Fisica Cosmica, San Miguel, Argentina.) Note the wide wings of H_ϵ, the broad H and K lines and the emission in many other lines.

Hα flare areas range from the frequent, very small brightenings of only a few millionths of the hemisphere to the rare flares of great extent that cover more than 1200 millionths of the hemisphere. The area of flares provides the basis for a simple system of 'importance' designation which indicates the general magnitude of the Hα event.

The frequency of flares varies widely with phase of the sunspot cycle. For solar cycle 20 the number of flares per year, of importance $\geqslant 1$, is shown in Table 9.1. It should be noted that even in the declining years of the 11-yr cycle, there can be intervals of flare abundance coincident with the transit of the then rare, but sometimes present, major centres of activity. *Subflares* (area $<$ 100 millionths) are $\approx$ 10 times as numerous as flares of imp. $\geqslant 1$. Small subflares occur with considerable frequency whenever there is a centre of activity, even in the years of solar minimum, and in the absence of flares of importance $\geqslant 1$.

TABLE 9.1
Number of flares per year, importance ⩾ 1, for solar cycle 20, 1964-1974.

Date	Imp. 1	Imp. ⩾ 2	Total	Yearly mean Zürich sunspot numbers
1964	11	2	13[a]	10.2
1965	82	6	88[a]	15.1
1966	240	50	290[a]	47.0
1967	458	63	521[a]	93.8
1968	518	51	569[b]	105.9
1969	567	64	631[b]	105.5
1970	585	52	637[b]	104.5
1971	176	13	189[b]	66.2
1972	218	17	235[b]	68.9
1973	128	13	141[b]	38.0
1974	150	13	163[b]	34.5

[a] = QBSA data reevaluated on basis of patrol times. See Reports UAG-2 and 19 (World Data Center A).
[b] = QBSA data.

9.3. Flare Area

The area of a flare is defined as the area of all parts of the chromosphere that have brightened in Hα. It is not limited to the area that has attained some specified intensity. The area for importance evaluation is measured at the time of maximum intensity, not at the time of maximum area which often occurs later in the flare. Area is reported either in millionths of the solar hemisphere (100 millionths = 3.04×10^8 km^2 on the solar surface) or in heliographic square degrees at the centre of the solar disk (1 square degree = 1.48×10^8 km^2). These are *'corrected areas'* derived from *'measured'* or *'apparent'* (observed) *areas* by applying a secant correction factor which takes foreshortening into account. However, this correction is not appropriate for flares more than 65° from the centre of the disk since flares have some extension in height (see limb flares). Apparent (observed) areas are strongly influenced by observing conditions (atmospheric seeing, exposure time) in particular on the small-scale photographs used in flare patrol. Areas reported for the same flare from different observatories may therefore vary considerably.

9.4. Flare Importance or Flare Class

The importance of a flare is an attempted evaluation of the magnitude of the Hα flare at the time of maximum intensity. Area and intensity are evaluated separately. There are five areal categories defined by the following limits: (see top of page 84).

The intensity is indicated by the letters *F*, *N*, and *B* for faint, normal, and bright, respectively, but there is no quantitative definition for the intensities corresponding to these letters. The combined area and intensity indices give dual flare importances or flare classes such as 1*N*, 2*F*, *SB*, etc.

Areal importance	Area 'millionths'	sq. degrees
S	<100	<2.0
1	100–250	2.1–5.1
2	250–600	5.2–12.4
3	600–1200	12.5–24.7
4	>1200	>24.7

Flares of importance *S* are called *subflares*, those of importance ⩾ 2 are referred to as *major flares.*

Before January 1, 1966 importance classes 1, 2 and 3 only were used. Subflares were designated by 1–, the largest flares (now class 4) by 3+. The + and – signs were used in addition to characterize flares of abnormally high and low Hα intensity, respectively.

9.5. Comprehensive Flare Index (CFI)

A comprehensive flare index, based on the ionizing, Hα, and radio frequency emissions of flares, was introduced by Dodson and Hedeman as an experimental effort to evaluate the electromagnetic radiation of the complex flare event. It is defined as:

$$(\mathrm{CFI}) = A + B + C + D + E,$$

where

A = Importance of ionizing radiation as indicated by the importance of associated SID; scale 1–3;

B = importance of Hα flare; scale 1–3 (3 stands for classes 3 and 4);

C = Log of 10 cm radio flux, in units 10^{-22} W m^{-2} Hz^{-1};

D = Effects in dynamic radio spectrum: Type II burst = 1, continuum = 2, Type IV burst = 3;

E = Log of 200 MHz flux in the same units as C.

These five components, when taken sequentially, constitute a crude profile of the radiation of the flare from short to long wavelengths. The value of the Index ranges from 0 to ~ 17 and values > 10 indicate flares with unusually strong electromagnetic radiation. The index is especially useful in the identification of the outstanding complex flare events and has been derived for the principal 'major' flares for the years 1955–1969 and 1970–1974.

Dodson, H. W. and Hedeman, E. R.: 1971, WDC-A Report UAG-14.
Dodson, H. W. and Hedeman, E. R.: 1975. WDC-A Report UAG-52.

9.6. Flash Phase

This is the period of rapid brightening and expansion of the flare to its maximum (Figure 9.2); it lasts from several to 15 min. Flares may start immediately with the flash or there may be some slow brightening preceding it, which is called *preheating* or *preflare*. Flares of low maximum brightness usually have no flash phase.

The flash phase frequently includes an *explosive* or *impulsive* phase which is a very sudden increase in brightness (within ≈ 1 min) of a small part of the flare (= *impulsive flare*) which sometimes is accompanied by rapid expansion. In general, 'energetic' flare processes such as impulsive microwave bursts, and hard X-ray bursts coincide with the optical impulsive phase. Flare kernels probably coincide with or are a part of the impulsive flare. The explosive flare seems to be the origin of flare sprays and Moreton waves.

9.7. Flare Kernels

These are small flare regions which brighten rapidly close to the magnetic inversion line and on both sides of it. They are extremely bright in Hα (with a very wide line profile) and in other optical emissions (He lines, metal lines) as well as in soft X-rays (*X-ray kernels*). They have also been called *hot cores.* White-light flare knots appear to coincide with flare kernels. They have been identified as the source of the impulsive hard X-ray bursts which indicates that they are the regions where particle acceleration occurs.

Vorpahl, J.: 1972, *Solar Phys.* **26**, 397.

9.8. Plage Flare

This term was suggested by McKenna-Lawlor to refer to the type of flare that is an impulsive brightening of the plage network. Such flares 'retain their positions relative to other features within the plage for the duration of the activity'. Tandberg-Hanssen suggested the same phrase for the type of flare that is 'an excess brightening of the pre-existing plage'. It seems probable that the same type of flare is intended by Zirin's (1974) use of the terms *in situ* or *confined* flares. It also seems probable that the term is used in contradistinction to events described as 'prominence flares'.

McKenna-Lawlor, S. M. P.: 1968, *Astrophys. J.* **153**, 367.
Tandberg-Hanssen, E.: 1973, *Earth Extraterrest. Sci.* **2**, 89.
Zirin, H.: 1974, *Vistas Astron.* **16**, 1.

9.9. Two-Ribbon Flare

Many flares develop as a pair of bright strands on either side of the main inversion line of the magnetic field of the active region. These are called two-ribbon or *two-strand flares.* A filament occupying the inversion line usually disappears at flare onset (Figure 9.2). In active regions with spots, however, two-ribbon flares occur also without a pre-existing filament. This applies, for example, to the *configuration A flare* whose ribbons cover two chains of spots of opposite polarity in close proximity.

The two flare strands often move apart with a velocity of 10–2 km s^{-1} and frequently are seen to be connected by bright or dark loops (flare loops) forming a tunnel or arcade.

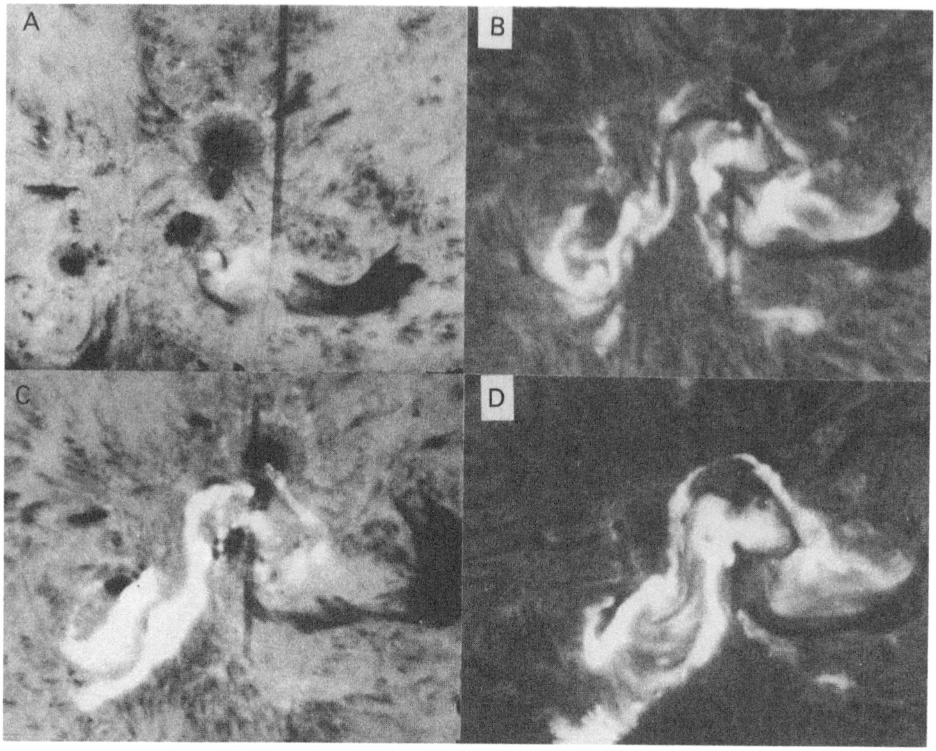

Fig. 9.2. Development of a typical two-ribbon flare with filament activation, filament disappearance and post flare loop system. (A) 1313 UT, Hα -0.75 Å, *S* surges, *F* active filament; (B) 1321 UT, Hα $+0.25$ Å, flare flash along the magnetic inversion line; (C) 1329 UT, Hα -0.75 Å, flare maximum, ribbons along the magnetic inversion line; (D) 1349 UT, Hα $+0.25$ Å, post-maximum phase with separated ribbons connected by flare loops seen in emission and absorption. (Fraunhofer Institut, Anacapri Observatory, June 15, 1972.)

Configuration A flares are energetic flares accelerating particles to high energies (proton flares).

Avignon, Y., Martres, M. J., and Pick, M.: 1963, *Ann. Astrophys.* **27**, 23.

9.10. Filament-Associated Flares

The disappearance (ascending, eruption) of a filament in or close to a spot group sometimes results in the brightening of a two-ribbon flare (Figure 9.3). Such prominence-associated flares include some of the largest and most energetic flares on record. If a quiescent filament, outside active regions with spots, disappears ('disparition brusque') sometimes chromospheric flare phenomena occur close to its former position; they range from a chain of moderately bright knots on one or both sides of the filament channel (*flare-like brightenings*) to large flares of the two-ribbon type (*prominence flares*).

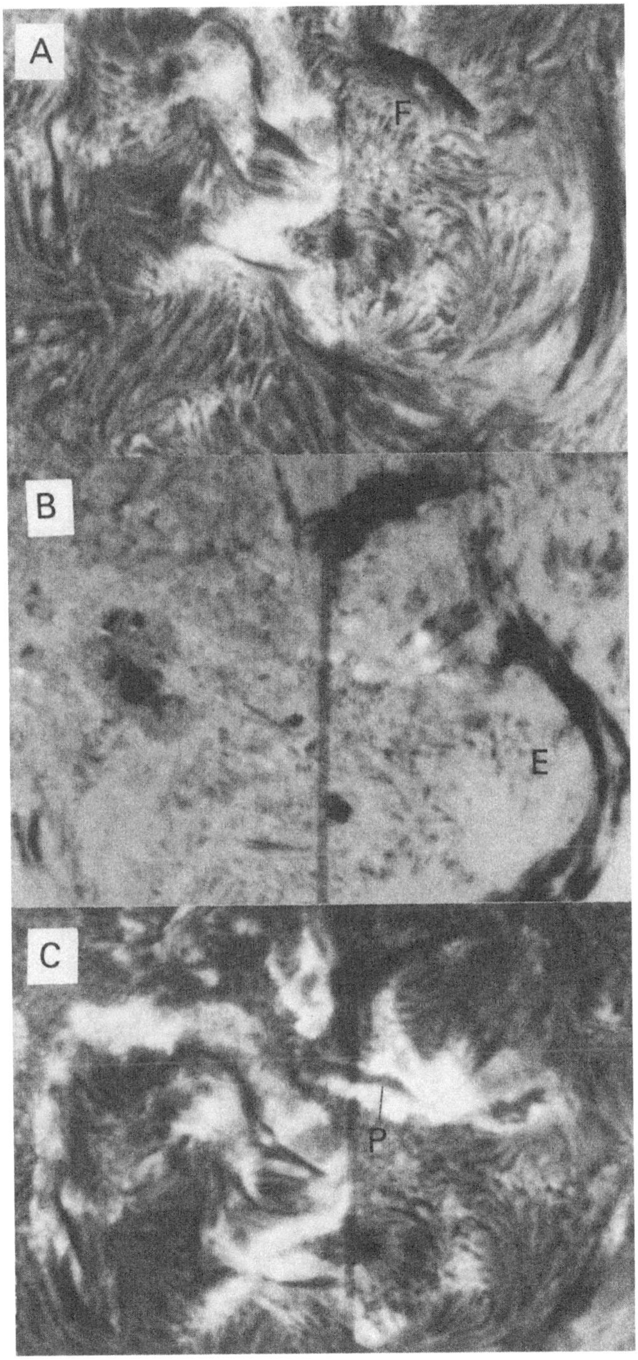

Fig. 9.3. Small flare associated with filament ascendance, June 15, 1972; (A) 0841 UT, small Hα filament *F*, slightly active; (B) and (C) 0930 UT, Hα and Hα −0.75 Å, flare encloses former position of filament *F* which is now ascending; *E* is another, possibly flare-activated filament. (Fraunhofer Institut, Anacapri Observatory.)

In general, major filament-associated flares in regions with small or no spots develop relatively slowly ('slow' flares); they rise to maximum intensity within 30–60 min and have a duration of several hours. They may attain large areas but usually show a relatively low maximum intensity (Figure 9.4). Associated X-ray and microwave emission is, corresponding to Hα, of the long-lived 'gradual rise and fall' type. Although these flares do not generally produce energetic particles or geomagnetic storms, there are a number of well-established exceptions to this statement.

Hyder revived Waldmeier's suggestion that the impact of material of the erupting prominence (filament) returning to the chromosphere produces the flare brightening (*Hyder's* or *infall-impact mechanism*).

Dodson, H. W. and Hedeman, E. R.: 1970, *Solar Phys.* **13**, 401.
Hyder, C.: 1967, *Solar Phys.* **2**, 49 and 267.
Zirin, H.: 1974, *Vistas Astron.* **16**, 1.

9.11. Homologous Flares

Sometimes successive flares occur in an active region at the same position and show a strikingly similar pattern of structure and development. Such series of recurring, nearly identical flares are called homologous flares. Homology has been found also in the shape of successive radio bursts. It has been suggested that the occurrence of homologous flares indicates that the basic circumstances determining flare occurrence persist through the series of flares; that is, that they are not changed or destroyed by the individual flares, not even by very large ones.

Ellison, M. A., McKenna, S. M. P., and Reid, J. A.: 1960. *Dunsink Obs. Pub.* **1**, 3.
Fokker, A. D.: 1967, *Solar Phys.* **2**, 316.
Zirin, H.: 1974, *Vistas Astron.* **16**, 1.

9.12. Limb Flares

Hα flares – or parts of them – extend into the corona up to heights $\gtrsim 10\,000$ km and thus are still visible at or even a little behind the solar limb as limb flares. The higher flare parts which usually appear as bright cones or mounds have physical properties different from those of the lower parts visible on the solar disk: electron densities are lower by an order of magnitude, and electron temperatures are higher by about a factor two, $n_e \approx 10^{11}–10^{12}$ cm^{-3}, $T_e \approx 15\,000$ K (Balmer lines). It is sometimes difficult to decide at first sight whether an observed limb feature is a flare or a bright flare-associated active prominence. Suitable criteria for a differentiation are provided by their development and motions. While flares, apart from expansion, are rather a static phenomenon, active prominences show fast, large-scale motions. There are, however, transitional types between flare and active prominence, e.g. the very first stage of a loop prominence system or of a flare spray. Other criteria are based on spectral characteristics (Figure 9.5).

9.13. Flare-Associated Phenomena

This term refers to a number of optically observed chromospheric and coronal phenomena which, in different ways, are related to flare occurrence.

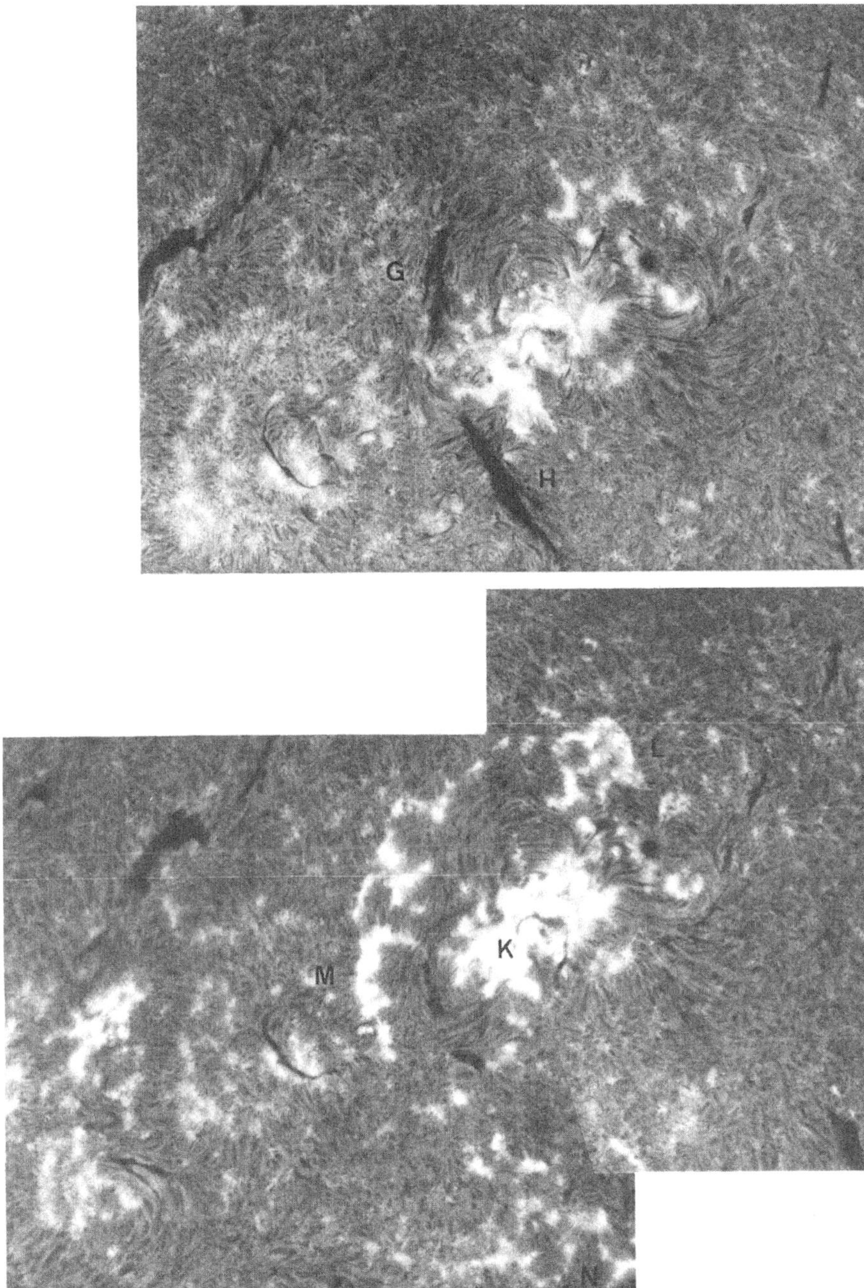

Fig. 9.4. Huge slow flare following the disappearance (disparition brusque) of activated filaments June 15, 1972, 1240–2100 UT (flare max. 1320 UT); top frame showing active filaments *G* and *H* and another flare in the active region 1030 UT; bottom frame showing slow flare consisting of a central part (*K*) in the active region and of brightenings forming a huge arc ($l \approx 500\,000$ km) outside the former position of the filaments (*L* – *M* – *N*). (Fraunh er Institut, Anacapri Observatory.)

Fig. 9.5. Hα limb flare, W-limb, December 2, 1967, 2247 UT. (Big Bear Solar Observatory.)

These are:

(1) *Filament activations* (see: section Prominences). There are three types of flare-associated filament activations:

(a) The pre-flare filament activation begins minutes or several tens of minutes before the flare and ends with the filament disappearing at flare onset. Chromospheric flare and/or an X-ray (coronal) brightening develop along the filament channel. This pre-flare activation is believed to be due to – and thus indicative of – pre-flare changes in the surrounding magnetic field which supply the flare energy (Figure 9.3).

(b) Filaments in the vicinity of flares are activated and frequently disrupted during flash or maximum phase. Changes in the magnetic field accompanying the flare or a shock wave produced by the flare may be responsible (Figure 9.3).

(c) Distant filaments start oscillating (see 'winking filaments') at the passage of a wave disturbance (see 'Moreton wave') originating in the flare.

(2) *Flare ejections* are (a) surges, (b) sprays, (c) fast ejections (see section 'Prominences'). They are (flare) material ejected during the flash or maximum phase.

(3) *Post-flare loops* develop out of energetic flares at about maximum. They are conspicuous in chromospheric and in coronal lines (optical, EUV, X-ray).

(4) Some types of coronal transients (see 'Corona').

(5) Moreton wave.

Bruzek, A.: 1974, in G. Newkirk (ed.), 'Coronal Disturbances', *IAU Symp.* 57, 323.

9.14. Moreton Wave

This wave disturbance – also referred to as *flare blast* or *flare wave* is generated by large flares probably at the explosive phase and propagates horizontally within a sector of angle of $\approx 90°$ with typical velocity of 1000 km s^{-1}. It manifests itself by:

(a) A bright and/or dark front moving in the chromosphere away from the flare. It is interpreted as the progressive depression of chromospheric elements along the front produced by a passing coronal pressure wave (*'sweeping skirt model'*) with subsequent return of the chromosphere to its original state (down-up motion of the chromosphere) (Figure 9.6).

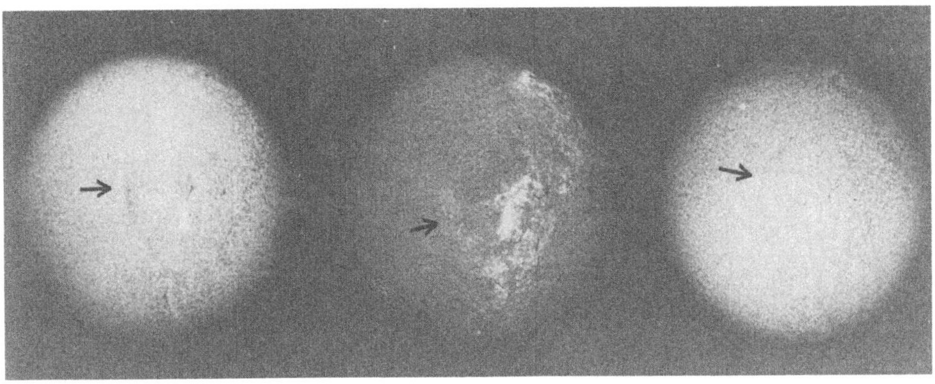

Fig. 9.6. Moreton wave associated with 3B proton flare August 28, 1966. It appears as bright front in Hα centre (centre frame) and as dark front in both Hα wings; 1531 UT. (Sacramento Peak Observatory.)

(b) oscillatory motions of distant filaments ('winking') which are excited by the passing coronal wave.

Smith, S. F. and Harvey, K. L.: 1971, in: C. Macris (ed.), *Physics of the Solar Corona*, Reidel, Dordrecht, p. 156.
Uchida, Y., Altschuler, M. D., and Newkirk, G. J.: 1973, *Solar Phys.* **28**, 495.

9.15. White-Light Flare

In rare cases, small parts in a flare become visible in white light for about 10 min during the flash phase; their spectrum is described as blueish. White-light flares observed on the central parts of the solar disk consist of one or two bright points or small areas which lie equidistant from the magnetic inversion line close to or in the penumbras of spots of opposite magnetic polarity. These represent conjugate points of a magnetic field loop overlying the flaring region. Maximum brightness is $\approx 50\%$ above photospheric intensity. The total energy emitted in the continuum is $\approx 10^{30}$ erg (Figure 9.7).

The very close time association of white light emission with hard X-ray and microwave bursts indicates an origin associated with accelerated particles. It is believed that beams of energetic electrons and/or protons bombarding the lower solar atmosphere produce the continuous emission.

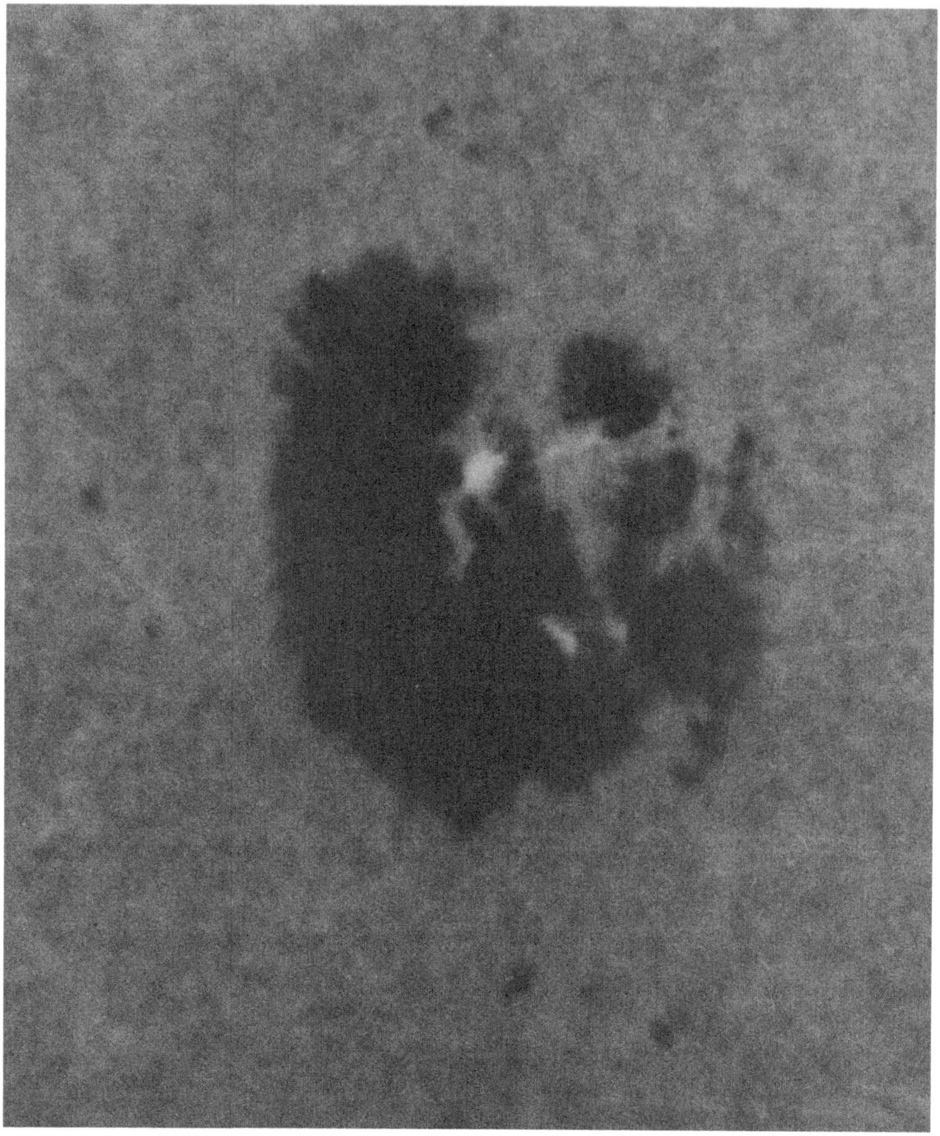

Fig. 9.7. White-light flare during 3B flare August 7, 1972; 1520 UT. (Sacramento Peak Observatory.)

White-light flares near the solar limb are larger and have the form of a bright facula. There the formation of negative hydrogen ions (at $n_e > 10^{15}$ cm^{-3}) may contribute continuous emission.

McIntosh, P. S. and Donnelly, R. F.: 1972, *Solar Phys.* **23**, 444.
Rust, D. M. and Hegwer, F.: 1975, *Solar Phys.* **40**, 141.
Švestka, Z.: 1970, *Solar Phys.* **13**, 471.

9.16. X-Ray Flare

The coronal regions affected by the flare emit soft X-rays ($\lambda > 1$ Å). Their spectrum includes continuum radiation (thermal bremsstrahlung and recombination continua) and emission lines of highly ionized atoms. The temperature of the source is up to $T_e = 4 \times 10^7$ K, its emission measure 10^{48}–10^{49} cm^{-3}. Electron densities are difficult to evaluate from X-ray measurements, but optical observations indicate $n_e \approx 10^9 - < 10^{11}$ cm^{-3}. Well-resolved photographs of soft X-ray flares have been taken by grazing-incidence X-ray telescopes on board OSO-7 and Skylab satellites. They show the X-ray flare with a bright *flare core* overlying the magnetic inversion line surrounded by a region of fainter, diffuse emission. The flare core contains complex loop systems and bright knots. During the decay phase loop systems are formed at successively increasing heights which finally lead to the large loops observed in the post flare phase.

The spatially unresolved, temporarily enhanced X-ray emission is referred to as the *X-ray burst*. Bursts have been, and still are, measured on a routine basis by a number of satellites in various wavelength bands, e.g. 0.5–3.0 Å, 1–8 Å, 8–20 Å. The time-intensity profile of soft X-ray bursts is similar to that of the Hα profile with some slight shifts of start, maximum and end of emission.

Very long-lived soft X-ray bursts (lifetime several hours) are associated with erupting prominences (filament disappearance).

X-ray flare importance or *class* is defined by the order of magnitude of the peak burst intensity, measured at the Earth in the 1 – 8 Å band, as follows

Importance	I_{max}
C	$1-9 \times 10^{-3}$ erg cm^{-2} s^{-1}
M	$1-9 \times 10^{-2}$ erg cm^{-2} s^{-1}
X	$\geqslant 10^{-1}$ erg cm^{-2} s^{-1}

The exact intensity is indicated by e.g.: M8 = 8×10^{-2} erg cm^{-2} s^{-1} or X5 = 5×10^{-1} erg cm^{-2} s^{-1}.

Pallavicini, R., Vaiana, G. S., Kahler, S. W., and Krieger, A. S.: 1975, *Solar Phys.* **45**, 411.
Švestka, Z.: 1976, *Solar Flares*, D. Reidel Publ. Co., p. 108.
Vorpahl, J. A., Gibson, E. G., Landecher, P. B., McKenzie, D. L., and Underwood, J. H.: 1975, *Solar Phys.* **45**, 199.

9.17. Hard X-Ray Flare

Hard X-ray ($E \geqslant 20$ keV) emission from flares occurs as short-lived impulsive bursts of non-thermal bremsstrahlung produced by streams of energetic electrons in the premaximum phase of the Hα and soft X-ray flare. Lifetimes are tens of seconds to minutes. Peak flux measured at the Earth is 10^{-6}–10^{-5} erg cm^{-2} s^{-1} above 10 keV. The spectrum is of the form

$$\frac{dI}{dE} = CE^{-\gamma}, \quad 2.7 \leqslant \gamma \leqslant 4.5 \quad \text{for} \quad 10 \text{ keV} < E \lesssim 70 \text{ keV}.$$

The source of the impulsive hard X-ray bursts is much smaller than the soft X-ray or Hα flare, and appears to coincide with intense Hα kernels (size, 10^8 km) brightening during burst occurrence. Below $E = 50$ keV there is also a longer-lived thermal component.

There is a strikingly good time correlation between impulsive X-ray bursts and impulsive microwave bursts even in fine details of the intensity curves. The correlation is best for very high frequencies.

Kane, S. R.: 1972, *Solar Phys.* **27**, 174.
Kane, S. R. and Lin, R. D.: 1972, *Solar Phys.* **23**, 457.
Vorpahl, S.: 1972, *Solar Phys.* **26**, 397.

9.18. EUV Bursts

EUV bursts ($250 < \lambda < 1350$ Å) occur in close time association with hard X-ray bursts and impulsive microwave bursts, but have a longer lifetime (typically ≈ 7 min). They appear in many coronal and transition lines during the optical flash phase. They are detected either by satellite-borne equipment or through their ionospheric effects. In general, EUV flares are found cospatial with the associated Hα flare. The total EUV peak flux measured at the Earth ranges from 10^{-2} to 10 erg cm^{-2} s^{-1}.

Many EUV bursts occur without an Hα or X-ray flare, indicating the possible existence of very small, flare-like instabilities (= *microflares*) that can only be observed through their effects in the transition region.

Švestka, Z.: 1976, *Solar Flares*, D. Reidel Publ. Co., p. 161.
Wood, A. T., Jr., Noyes, R. W., Dupree, A. K., Huber, M. C. E., Parkinson, W. H., Reeves, E. M., and Withbroe, G. T.: 1972, *Solar Phys.* **24**, 169.

9.19. Flare Gamma Ray Emission

Flare gamma ray emission was observed by the OSO-7 satellite during the large flares of Aug 4 and 7, 1972.

Between 0.36 and 7 MeV a continuum was detected for the Aug 4, 3B flare whose spectrum followed a power law with index $\gamma = 3.4$ between 360–700 keV and an exponential law with $E_0 = 1.0$ MeV for $E > 700$ keV. It is suggested that the radiation is non-thermal bremsstrahlung from electrons with energies > 100 keV.

Gamma ray lines were detected at 0.5 MeV corresponding to the electron positron annihilation energy, at 2.2 MeV corresponding to neutron capture by protons in the photosphere, and at 1.17, 1.33, 4.4, and 6.1 MeV which are due to transitions in excited nuclei.

Chupp, E. L., Forrest, D. J., and Suri, A. N.: 1975, in S. R. Kane (ed.), 'Solar Gamma-, X-, and EUV Radiation', *IAU Symp.* **68**, 341.
Suri, A. N., Chupp, E. L., Forrest, D. J., and Reppin, C.: 1975, *Solar Phys.* **43**, 415.

9.20. Particle Flare or Energetic Flare

The most energetic flares are the very rare *cosmic ray flares* which emit protons

> 500 MeV which – arriving at the Earth $\gtrsim 15$ min after flare flash onset – produce a *Ground Level Effect* (GLE). Only 20 GLE's have been recorded between 1942–1972. The largest energy measured was 15 GeV in the flare of 23 February 1956.

Proton flares emit protons, 10 MeV $\lesssim E < 100$ MeV, which produce anomalous ionisation in the *D* layer of the polar ionosphere and thereby cause *Polar Cap Absorption* (PCA). Cosmic ray flares and the majority of proton flares are two-ribbon flares, many of them with loop prominence systems. They are associated with strong Type IV and microwave bursts and hard X-ray bursts. The number of proton flares in the period 1955–1969 was about 380. In addition, there were many other particle events with ambiguous flare sources or with possible association with flare events on the invisible hemisphere: 25–30% of the PCA's could be attributed to flares occurring on the invisible hemisphere up to 40° beyond the W solar limb (*invisible flare*).

The maximum *radiation dosage* from a flare was found in one case about 60R hr^{-1} behind 0.22 g cm^{-1} Al. The total dosage from the large flare of 12 November 1960 was estimated at 700R.

All solar proton events are accompanied by non-relativistic electrons ($E > 40$ keV, fluxes up to ≈ 5000 electrons cm^{-2} s^{-1} ster^{-1}). However, the majority of electron events (electron flux <100 cm^{-2} s^{-1} ster^{-1}) occur without detectable protons. The associated flare is called an *electron flare*. More than 80% of electron flares are associated with Type III and microwave bursts; usually a hard X-ray burst is also present.

Relativistic electrons from flares ($E = 3$ MeV – > 12 eV) have been recorded in a number of cases in recent years. They are generally accompanied by a proton event.

Švestka, Z.: 1976, *Solar Flares*, D. Reidel Publ. Co., Dordrecht, p. 238.
Švestka, Z. and Simon, P. (eds): 1975, *Catalogue of Solar Particle Events, (1955–1969)*, D. Reidel Publ. Co., Dordrecht.

9.21. Flare Mechanisms

No satisfactory theory of the entire flare event exists as yet. Most descriptions are heuristic, or at best semi-quantitative, and there is no general agreement about even the physical mechanisms involved.

Considerations of energy and timescale point to the solar magnetic field as the immediate source of the flare. The field in the upper chromosphere and lower corona exists in a generally stable configuration that is close to being force-free almost everywhere. The disturbance of this state by the emergence of new flux through the photosphere or by the motion of the field footprints tied into the subphotospheric plasma can lead to a metastable situation when oppositely directed fields are squeezed together. A current sheet forms with a thickness of the order of the ion gyro-radius (at coronal temperatures) in which magnetic diffusion can operate and field annihilation take place. This is *Sweet's mechanism* and releases energy relatively slowly. It can perhaps explain the thermal build-up to the flash phase of flares.

The flash phase itself requires a sudden lowering of the electrical conductivity either by the current sheet meeting much lower temperature plasma, a thermal instability setting in or the onset of a turbulent (tearing-mode) instability. This increases the magnetic diffusion rate and the rate of energy release is then only limited as the accelerating flow bringing the flux into the sheet reduces the magnetic field strength. This is *Petschek's mechanism* and converts magnetic energy into kinetic energy, predominantly of electrons, which are accelerated by the electric field in the current sheet. These electrons will produce the hard X-ray bursts by bremsstrahlung but it is not clear whether they cause the optical flare kernels directly by impact in the chromosphere or indirectly by conduction along magnetic field lines or by shock wave heating. Indeed all three processes may play a role. Presumably heat then diffuses away from the kernels in the chromosphere to produce the low temperature flare. It is likely that the chromospheric heating evaporates some of the denser plasma of the low temperature flare which will fill the field lines delineating the flaring region in the lower corona. The acceleration of high energy ions is an unsolved problem and seems to call for a separate but contemporaneous mechanism. One workable model invokes a system of hydromagnetic shocks in the plasma above the flaring region. Intersecting shocks act preferentially like closing magnetic mirrors which accelerate charged particles trapped between them by the Fermi mechanism.

Canfield, R. C., Priest, E. R., and Rust, D. M.: 1974, in Y. Nakagawa and D. M. Rust, (eds.), *Flare Related Magnetic Field Dynamics*, HAO-NCAR, Boulder, p. 361.
Sonnerup, B. U. Ö.: 1973, in R.. Ramaty and R. G. Stone (eds.), *High Energy Phenomena on the Sun*, Goddard Space Flight Center, Greenbelt, p. 357.
Švestka, Z.: 1976, *Solar Flares*, D. Reidel Publ. Co., Dordrecht, p. 300.

10. PROMINENCES

E. TANDBERG-HANSSEN

10.1. Prominence

The term prominence is used for a large variety of objects which are characterized by their occurrence in the chromosphere and/or in the corona, by their greater density than their coronal surroundings and by temperatures ranging from about 10^4 K to coronal values at the interface to the corona. They appear as bright (cool and dense) features (in the corona) above the solar limb; when seen projected against the solar disk nearly all prominences show up in absorption as *dark filaments.*

Two large classes of prominences may be distinguished:

(1) *Quiescent prominences* are relatively stable structures with lifetimes of many months. In their well-developed stages they occur outside active regions with sunspots, and consist predominantly of low-temperature plasma with physical parameters as follows:

hydrogen density	$n_H \approx 10^{11}\,\text{cm}^{-3}$,
electron density	$n_e \approx 0.8\,n_H$,
electron temperature	$T_e \approx 7000$ K,
magnetic field	$B = 5 - 10$ G.

(2) *Active region prominences* comprise:

(a) the relatively stable *plage filaments* which run along a main magnetic inversion line in or at the border of an active region but often approach or even enter at one end a major sunspot. They are predecessors of quiescent prominences, and usually have phases of some activity.

(b) *active prominences* such as surges, sprays, and loops which show fast changes and violent motions. They have lifetimes of minutes to hours, an internal magnetic field ≈ 100 G and a larger fraction of their mass at high temperatures. Some are related to spots (spot prominences) or associated with flares, some may even be confused with flares (Figure 10.1).

Bruzek, A. and Kuperus, M.: 1972, *Solar Phys.* **24**, 3.
Martin, S. F.: 1973, *Solar Phys.* **31**, 3.
Martres, M. J., Michard, R., and Soru-Iscovici, I.: 1966, *Ann. Astrophys.* **29**, 249.
Tandberg-Hanssen, E.: 1974, *Solar Prominences*, D. Reidel, Dordrecht.

10.2. Prominence Classification

Prominences have been classified in several ways. Already Secchi (1875) used their degree of dynamic activity to divide them into quiescent and active (eruptive) prominences. Widely used is Pettit's classification (1925) in which the objects are placed in one of five

Bruzek and Durrant (eds.), Illustrated Glossary for Solar and Solar-Terrestrial Physics. 97–109.

Fig. 10.1. Active prominence photographed in the λ1032 Å line of O VI. (Center for Astrophysics, Cambridge, Mass.)

classes: active, eruptive, sunspot-related, tornado or quiescent. Both Severny (1950) and de Jager (1959) used mainly prominence motions as criteria for their classifications; de Jager dividing the objects into quiescent and moving prominences. A similar division is used by Zirin (1966) whose sharply defined classes depend on the relation to solar activity – in particular whether the prominences are related to flares or not. Those that are related to flares exhibit violent motions and are short-lived. Menzel and Evans (1953) introduced a two-dimensional classification based on (1) whether the prominence seems to originate from above (i.e., the corona) or from below, and (2) whether the prominence is associated with sunspots or not (i.e., associated with solar activity or not) (see also 'Prominence').

While the above-mentioned classifications all are morphological, Waldmeier (1949) and Zirin and Tandberg-Hanssen (1960), Tandberg-Hanssen (1963) have classified prominences according to relative intensities of spectral lines from objects seen in emission above the limb. In this way, one may distinguish between flares and different forms of bright, active prominences (surges, loops), as well as between active and quiescent prominences as defined from morphological studies.

Tandberg-Hanssen, E.: 1974, *Solar Prominences*, D. Reidel Publ. Co., Dordrecht, p. 5.

10.3. Prominence Phenomenology

Prominences are frequently described by their various shapes. Besides surge, loop – which are dealt with in separate entries – the following terms are used in the literature (most having been introduced by the Menzel-Evans classification):

Caps are seen in emission above the limb as bright, low-lying prominences near active regions and have lifetimes from hours to days. Surges frequently are ejected from the edges of caps. They may be the limb manifestation of fibrils and arch filaments (Figure 10.2).

Fig. 10.2. *Cap prominence* Hα, near spot group W-limb July 9, 1970. (Fraunhofer Institut, Anacapri Observatory.)

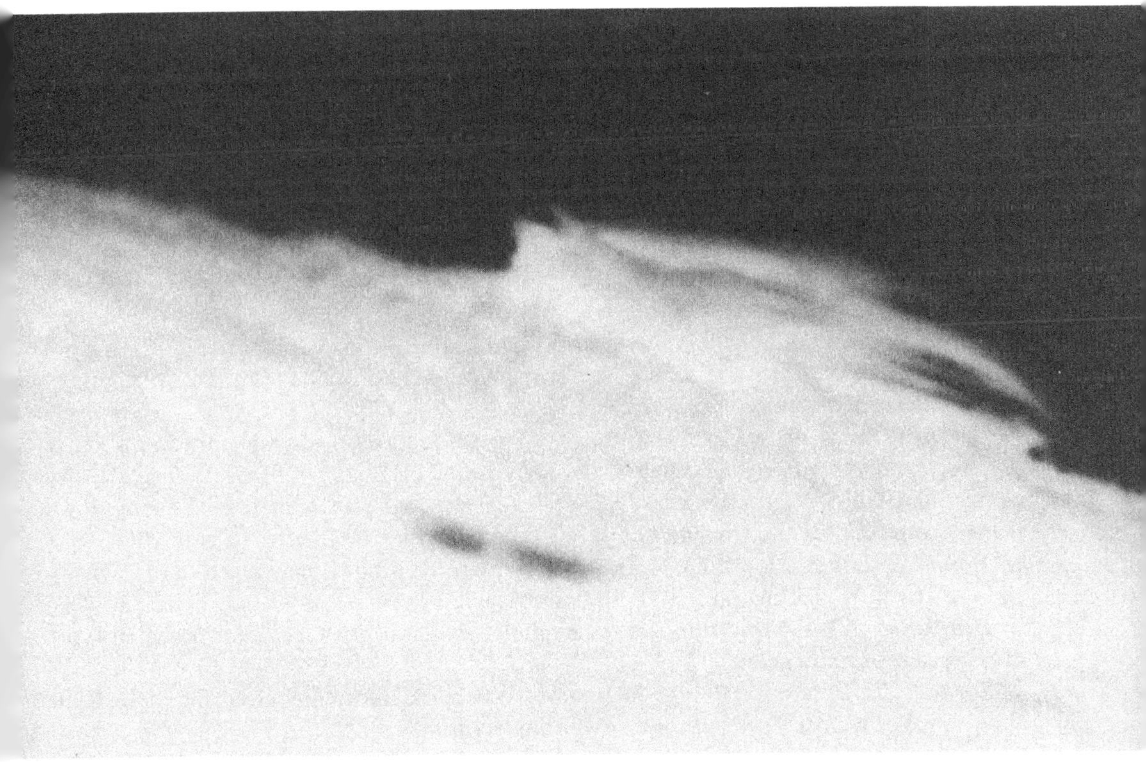

Fig. 10.3. *Coronal cloud* Hα. (Sacramento Peak Observatory, New Mexico.)

Coronal clouds are irregular objects suspended in the corona with matter streaming out of them into nearby active regions. The coronal clouds last for a day or more at heights of several tens of thousands of km (Figure 10.3).

Coronal rain is material flowing down along strongly curved trajectories into active regions, seemingly after condensing out of the corona, often at great heights. The velocity along these paths is 50–100 km s^{-1}. Similar to coronal rain is material flowing down from coronal clouds and from certain activated quiescent prominences (Figure 10.4).

Mound describes a low-lying prominence seen on the limb. It may reach flare intensity and is difficult to distinguish from limb flares.

Hedge Row, Tree, Tree Trunk are self-explanatory descriptive terms for the main types of quiescent prominences.

Tornado is a rare type in which a vertical spiral structure gives the object the appearance of a closely wound rope or whirling column.

Fountain is a distinctive type of ascending prominence which is characterized by closed-system transfer of chromospheric material along an arch or loop. The entire prominence envelope steadily rises and expands through the corona. Fountains appear to be well-contained by coronal magnetic fields.

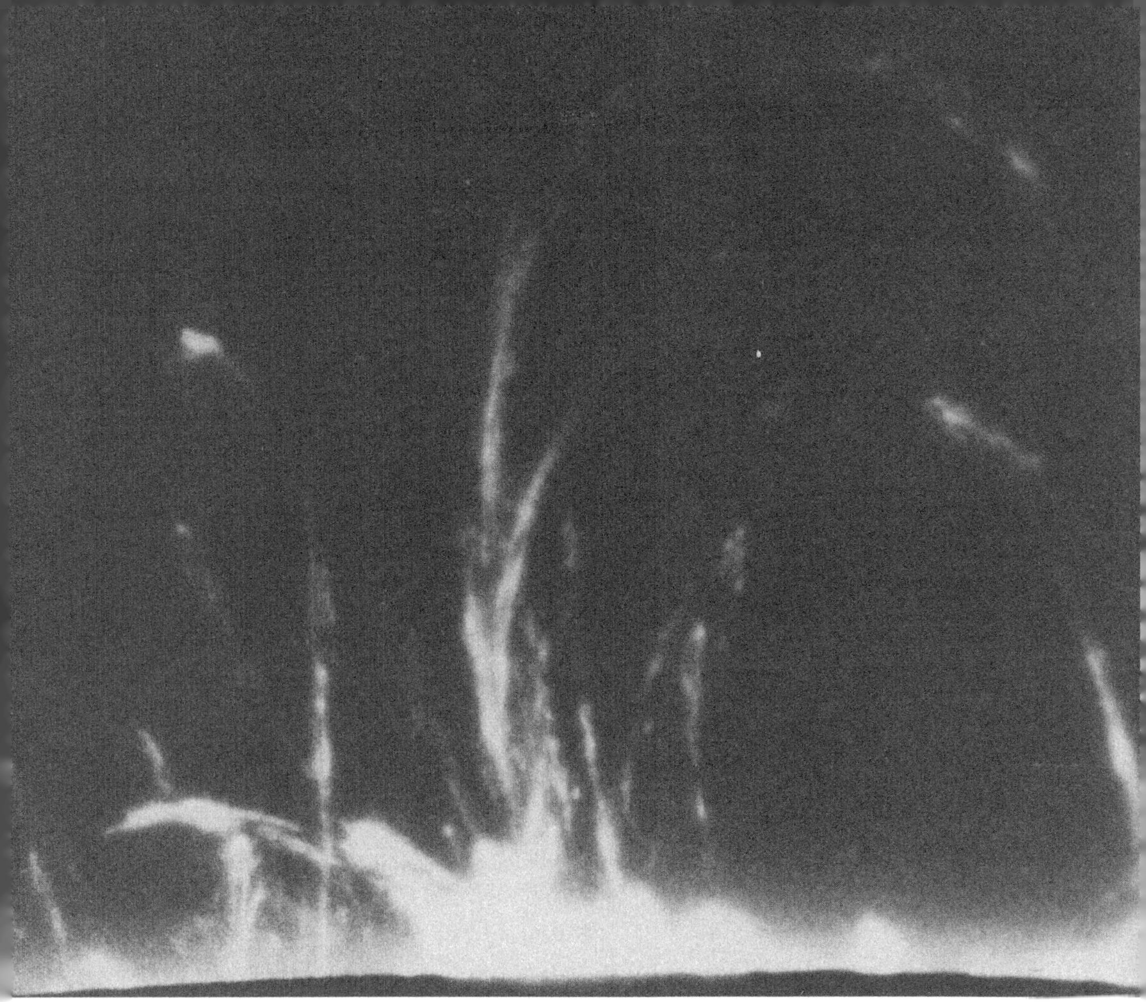

Fig. 10.4. *Coronal Rain* Hα. (Sacramento Peak Observatory, New Mexico.)

Tandberg-Hanssen, E.: 1974, *Solar Prominences*, D. Reidel Publ. Co., Dordrecht, Holland, p. 8, 29.
Tandberg Hanssen, E., Hansen, R. T., and Riddle, A. C.: 1975, *Solar Phys.* **44**, 417.

10.4. Quiescent Prominence

Quiescent prominences (filaments) are long, flat, sheet-like structures, nearly perpendicular to the solar surface; typical dimensions are: length 60–600 Mm, height 15–100 Mm, thickness 4–15 Mm. Well-developed specimens consist of a series of arches or trees whose feet are anchored in supergranulation borders. Except in 'activated phases' they show no large-scale motions, develop very slowly and have lifetimes of several months (Figure 10.5).

High-resolution Hα limb photographs reveal an internal fine structure of more or less vertical ropes of diameter $\lesssim 300$ km. Material is seen slowly streaming down these ropes ($v \approx 1$ km s^{-1}). The resulting mass loss is considerable and demands that material is continuously fed into the prominence in order to maintain it (Figure 10.6).

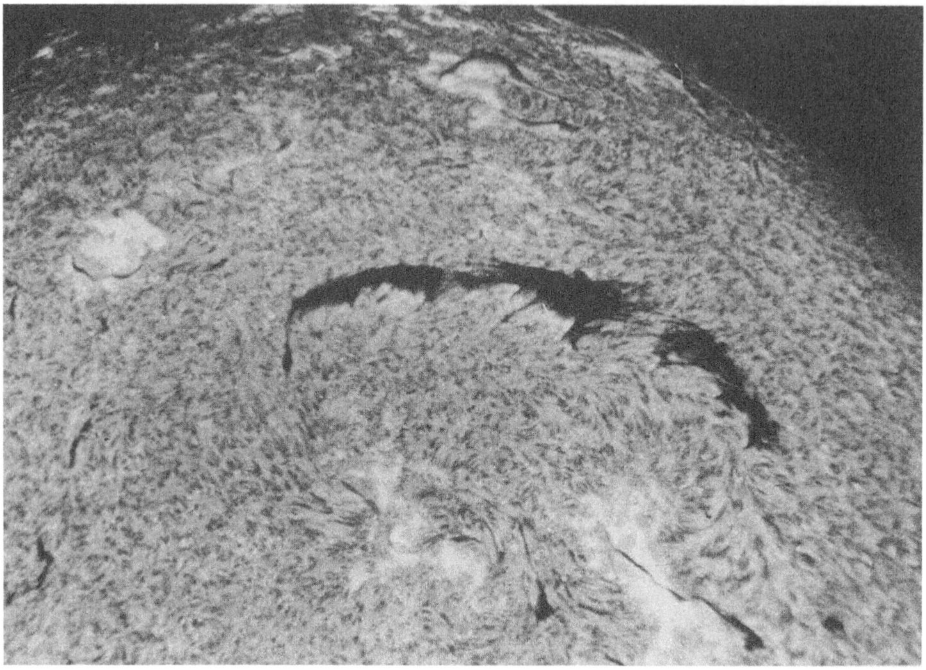

Fig. 10.5. *Quiescent Hα filament* (= quiescent prominence seen in absorption against the solar disk), January 31, 1959; length of the filament 450 000 km. (Fraunhofer Institut, Anacapri Observatory.)

Quiescent prominences are a typical feature of old, weak magnetic regions which may or may not still be accompanied by a weak plage. They are the mature and old stages of smaller, more active filaments which are primarily formed either (1) inside active regions with spots (*type A filament*), or (2) between two close active regions (*type B filament*) or (3) in the weak fields of remnants of active regions (see 'plage filaments').

Quiescent prominences disappear either by (1) slow dissolution, (2) flowing down into the chromosphere, or (3) eruption ≡ sudden disappearance (disparition brusque).

de Jager, C.: 1959 in S. Flügge (ed.), *Encyclopedia of Physics* **52**, 226.
Martin, S. F.: 1973, *Solar Phys.* **31**, 3.

10.5. Prominence Zones

The quiescent filaments/prominences developing out of active regions have a considerable extension in latitude and therefore become gradually more and more inclined through the action of the differential rotation while they slowly migrate to higher latitudes. There they form the two *royal zones of prominences* whose maxima occur ≈ 10° polewards of the spot zones and which migrate (parallel to the spot zones) towards the equator during the solar cycle.

Besides the royal zones there exist *polar zones of filaments/prominences* polewards of ≈ 40° lat. The polar filaments are oriented nearly parallel to the equator and sometimes

Fig. 10.6. *Quiescent prominence* showing predominantly vertical fine structure in Hα. (Sacramento Peak Observatory, New Mexico.)

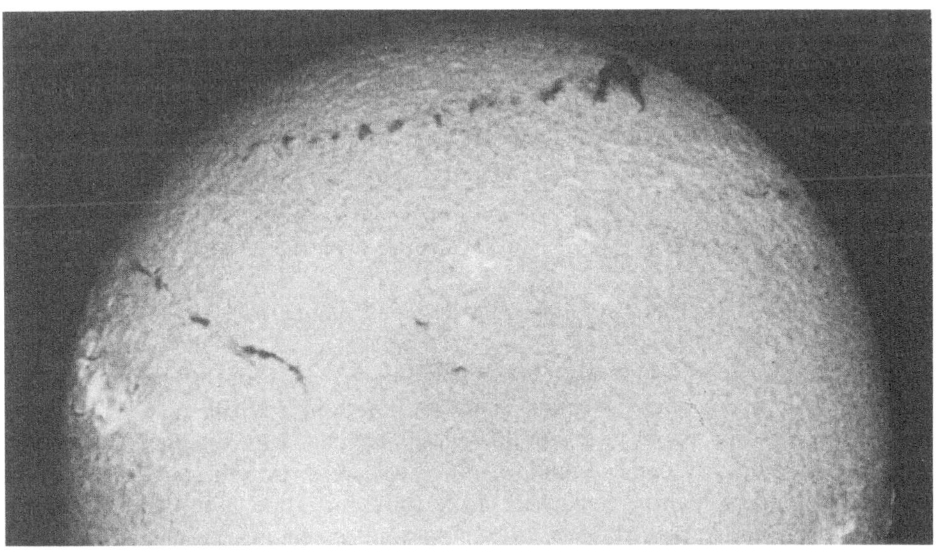

Fig. 10.7. Hα filtergram showing the *polar crown* and a quiescent filament in the royal prominence zone of the northern hemisphere, October 4, 1966. (Fraunhofer Institut, Anacapri Observatory.)

form a nearly uninterrupted *polar crown* around the polar caps (Figure 10.7). The polar crown is probably maintained by poleward migrating filaments originating at lower latitudes. The polar zones do not appear until ≈ 3 yr after spot maximum and migrate towards the pole, reaching it about the time of the next maximum. The migration of these filaments through the polar regions accompanies a reversal in polarity of the polar magnetic fields.

Waldmeier, M.: 1955, *Ergebnisse und Probleme der Sonnenforschung*, Leipzig, p. 254.
Waldmeier, M.: 1973, *Solar Phys.* **28**, 389.

10.6. Activated Prominences

Various types of large scale motions may occur in originally quiescent and stable, active region filaments/prominences; they may become activated filaments/prominences:

(a) Increased internal (turbulent or helical) motion or flow along the filament may occur which are associated with an enlargement and darkening of the filament or a brightening of the prominence. This activation may die away after ≈ 1 hr, or it may continue in an accelerated rising motion, i.e. in the production of an ascending (eruptive) filament/prominence. This type of activation indicates an instability of the supporting magnetic field which in active regions may lead to flare occurrence (see Figures 9.2, 9.3).

(b) Quiescent prominences sometimes show an active phase in which material flows away from the top region along an extended curved trajectory into a distant '*centre of attraction*'. This flow which occurs at a velocity of ≈ 100 km s^{-1} may stop after several hours or may continue until the whole prominence material has descended into the chromosphere. This is one way of quiescent prominence dissolution.

(c) *Winking filament*. This type of activation consists of a more or less vertical, damped oscillatory motion with periods between 6 and 40 min, lasting for 2–5 oscillations. It is excited by the passage of a MHD shock wave originating in a (sometimes rather distant) large flare and travelling with velocities in the range 400–1000 km s^{-1}. The winking impression is due to the Doppler effect which shifts the image of the filament periodically out of and into the narrow passband of Hα filters. (See Moreton wave).

Ramsey, H. E. and Smith, F. S.: 1966, *Astron. J.* **71**, 197.
Smith, F. S. and Ramsey, H. E.: 1964, *Z. Astrophys.* **60**, 1.

10.7. Ascending (Eruptive) Prominence

Activated prominences (quiescent and active region) occasionally become strongly destabilized and ascend with increasing velocities. Reaching $v > 100$ km s^{-1}, sometimes more than the escape velocity, the prominence material partly disappears high in the corona, partly returns in a helical motion along huge arches into the chromosphere. As a rule, the prominence/filament reappears in roughly the same shape and at the same position after some time. At least three types of ascending prominences may be distinguished:

(a) Old quiescent prominences far from active regions tend to ascend and to reappear after several days several times during their life. The rising motion starts at a few km s^{-1} but may attain high values until the prominence disappears after several hours. This is the classical *Disparition Brusque* (DB) or *Sudden Disappearance* (Figure 10.8). It may be followed after ≈ 1 hr by the slow brightening of flare-like points or even major flare strands along the base of the prominence (Figure 9.4). There are indications that some DB's are initiated by a disturbance emanating either from a newly emerging flux region or from a flare. In the majority of cases, however, no clear evidence for an external cause exists.

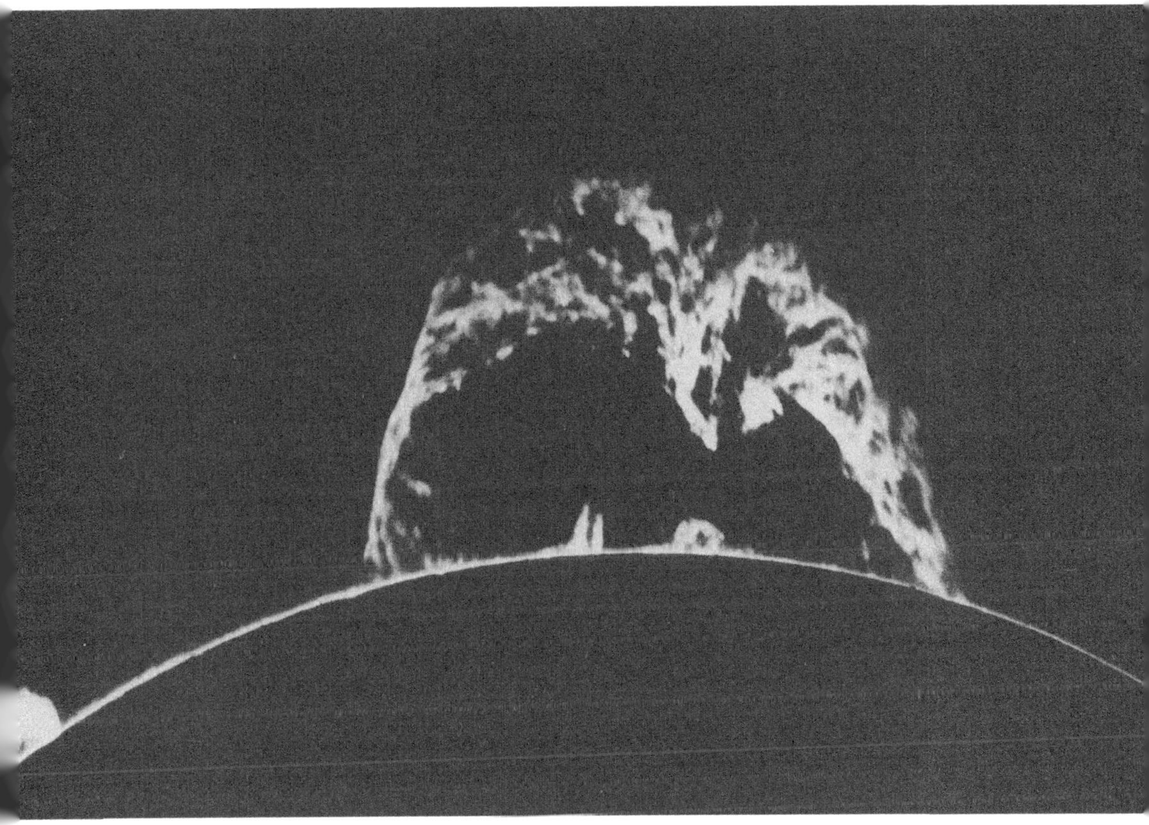

Fig. 10.8. *Ascending quiescent prominence*, March 13, 1970, 2011 UT. (Institute for Astronomy, Haleakala Observatory, Hawaii.)

(b) The same may happen to active region filaments, however with a considerably compressed time scale. The interval between first activation and disappearance is reduced to $\lesssim \frac{1}{2}$ hr; a flare brightens invariably while the filament is still ascending with high velocity, and the filament reappears after a few hours. This DB gives rise to the typical two-strand flare along the magnetic inversion line in active regions (DB-flare event) (Figures 9.2, 9.3).

(c) Filaments also become ascending and eruptive during the flash and maximum phase of nearby flares apparently by a direct flare action. To this class belongs the second type of sprays (prominence spray).

Bruzek, A.: 1974, in G. Newkirk (ed.), 'Coronal Disturbances', *IAU Symp.* **57**, 323.
Hyder, C.: 1967, *Solar Phys.* **2**, 49.
Smith, F. S. and Ramsey, H. E.: 1964, *Z. Astrophys.* **60**, 1.

10.8. Ejections

Some types of prominence consist of material ejected from the chromosphere, from a flare or a prominence. Best known are the surge- and spray-type prominences. A peculiar class of ejections are the rarely observed *fast ejections* which reach velocities $> 1000\ \mathrm{km\ s^{-1}}$ in a few minutes. They seem to consist of a compact portion of a flare which is ejected without fragmentation. It may be assumed that they occur much more frequently than they are observed since most of them will escape observation because of the large Doppler shift and short lifetime, both due to the high velocity.

Bruzek, A.: 1969 in C. de Jager and Z. Švestka (eds.), *Solar Flares and Space Research*, North Holland, Amsterdam, p. 6.

10.9. Surge

Surges are straight or slightly curved spikes which are shot out of a small luminous mound at velocities of 100–200 $\mathrm{km\ s^{-1}}$. They reach heights in the corona of up to 200 Mm and typically last 10 to 20 min. The surge material either fades or returns into the chromosphere along the trajectory of ascent. On the solar disk surges appear usually in absorption but, in their initial phase, sometimes in emission. They originate from small flare-like or bomb-like brightenings close to spots or pores. Many (small) surges start at penumbral borders and are directed radially away from the spot. Surges show a strong tendency to recur, at a rate $\approx 1/\mathrm{hr}$ for small surges (Figure 10.9).

The trajectory of the moving plasma and its collimation indicate that surges are confined by a more or less radial field. The magnetic field strength in surges is ≈ 50 G decreasing with height. Various mechanisms have been proposed for the initial acceleration of the surge material. There is observational evidence that the required energy may be provided by the underlying magnetic field. The mass and energy contained in large flare-associated surges is 10^{15}–10^{16} g (density, 10^{11}–$10^{12}\ \mathrm{cm^{-3}}$) and 10^{30} erg respectively; the energy of small, bomb-associated surges is 5×10^{27} erg.

Bruzek, A.: 1974, in G. Newkirk (ed.), 'Coronal Disturbances', *IAU Symp.* **57**, 323.
Roy, J. R.: 1973, *Solar Phys.* **28**, 95; **32**, 139.

10.10. Spray

Sprays are vigorous, flare-associated ejections of plasma which frequently appear disrupted into bright clumps. After an initial very high acceleration (a few $\mathrm{km\ s^{-2}}$) the

Fig. 10.9. Hα *surge* prominence seen in absorption and in emission projected against the solar disk, May 22, 1970. (Big Bear Solar Observatory.)

material follows slightly decelerated trajectories spreading over a large volume; maximum velocities are $\gtrsim 400$ km s^{-1}, frequently larger then the velocity of escape. A certain fraction of the material is seen to return to the solar surface, the rest fades or escapes from the Sun.

Two types of spray may be distinguished:

(1) the *flare spray*, or spray proper, which emanates from a rapid, explosion-like expansion of a flare or part of it. At the solar limb a very bright expanding mound appears which suddenly disrupts and expels the spray material. This development suggests that flare plasma originally constrained in a closed magnetic field configuration eventually bursts the field open due to its increasingly high kinetic energy density and escapes into the corona. Disk observations of sprays are scarce since the high speed plasma (giving large Doppler shifts) escapes normal observations close to the Hα centre (Figure 10.10).

(2) *prominence spray*: active region filaments sometimes erupt and are rapidly driven away as a spray during the flash or maximum phase of a nearby solar flare. This type of spray may be produced by a flare shock wave, but a stream of particles emanating from the flare has also been invoked.

Bruzek, A.: 1974, in G. Newkirk (ed.), 'Coronal Disturbances', *IAU Symp.* **57**, 323.
Smith, E. v. P.: 1968, in Y. Öhman (ed.), 'Mass Motions in Solar Flares', *Nobel Symp.* **9**, 137.

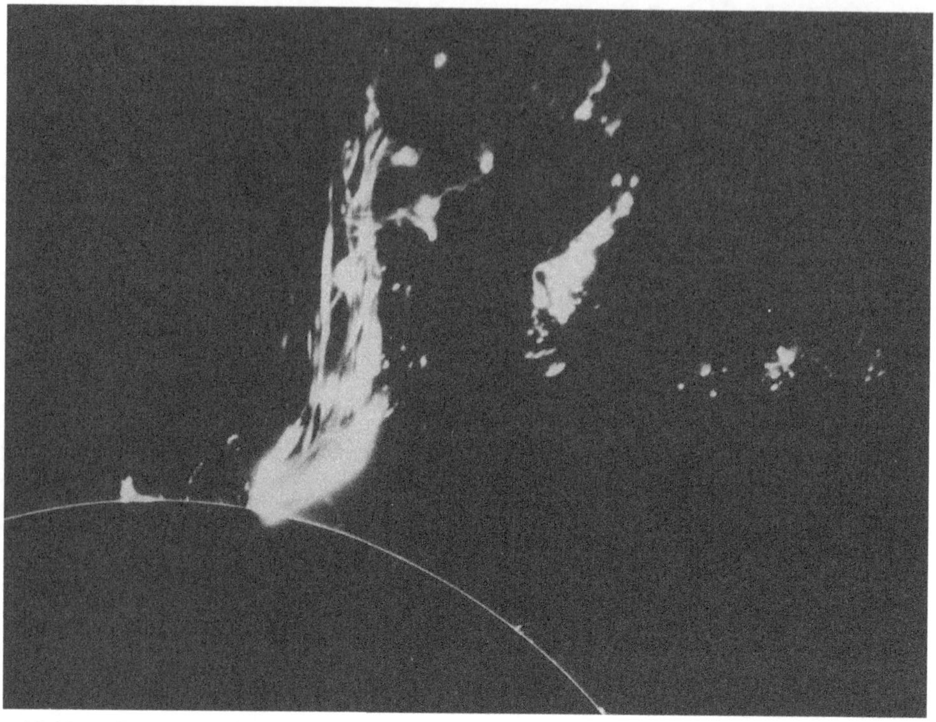

Fig. 10.10. *Spray* type prominence, January 3, 1969; height 600 000 km. (Institute for Astronomy, Haleakala Observatory, Hawaii.)

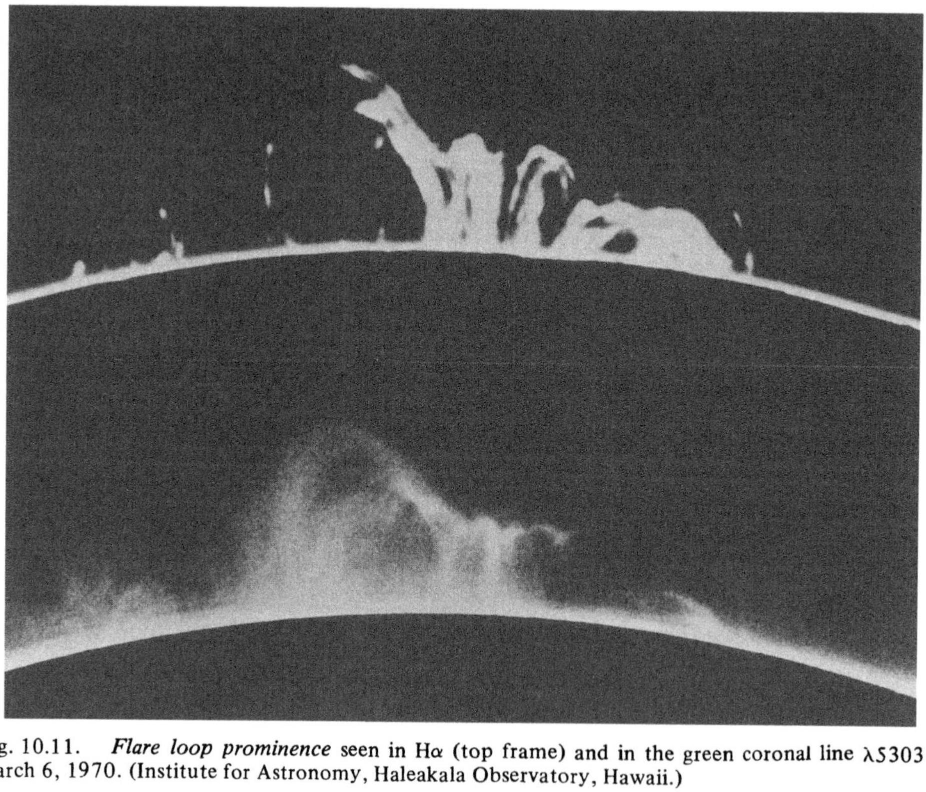

Fig. 10.11. *Flare loop prominence* seen in Hα (top frame) and in the green coronal line λ5303 Å, March 6, 1970. (Institute for Astronomy, Haleakala Observatory, Hawaii.)

10.11. Loop Prominences

Two types of loop-shaped prominence are observed:

(a) *single loops* with material ascending in one branch and descending in the other branch, usually into a sunspot. Velocities are ≈ 30 km s^{-1}, lifetime of loops $\simeq 15$ min, heights ≈ 50 Mm. These loops sometimes originate as surges.

(b) *flare loops* develop out of flares as *loop prominence systems (LPS)* bridging the magnetic inversion line. They form either a *loop tunnel* of parallel loops connecting two flare strands or a fan of loops converging into one or two spots. Mass motion is downwards along both branches with velocities up to 150 km s^{-1}. Lifetimes are several hours (see also *Solar Flares*) (Figure 10.11).

Bruzek, A.: 1964, *Astrophys. J.* **140**, 746.

11. SOLAR RADIO EMISSION

A. D. FOKKER

11.1. Quiet-Sun Radio Emission

Outside active regions the Sun produces radio emission that is due entirely to bremsstrahlung from thermal electrons – the quiet-Sun radio emission. Radiation of frequency f originates at and above the level where $f = f_p$ (= plasma frequency), i.e. meter wavelengths in the corona, decimeter wavelengths in the transition region and centimeter wavelengths in the chromosphere. The spectrum is shown in Figure 11.1.

The distribution of intensity over the solar disk depends on the distribution of electron density and temperature along the various ray trajectories. At centimeter wavelengths the intensity is rather uniform over the disk except for a sharp peak at the limb (*limb-brightening*). At decimeter wavelengths, a rather broad enhancement exists near the limb at small to moderate heliographic latitudes. At meter wavelengths, there is a gradual decline of intensity from the centre outwards.

The longer the wavelength the farther the *radio Sun* extends beyond the optical disk; at λ 3 m, the diameter is roughly twice the optical diameter. The brightness temperature of the quiet Sun varies from $\lesssim$ 5000 K in the mm-range to 10^6 K in the m-range. At sunspot maximum the quiet Sun intensity is roughly 25–60% greater than at the minimum of the solar cycle.

Kundu, M. R.: 1965, *Solar Radio Astronomy*, John Wiley, New York, Ch. 5.

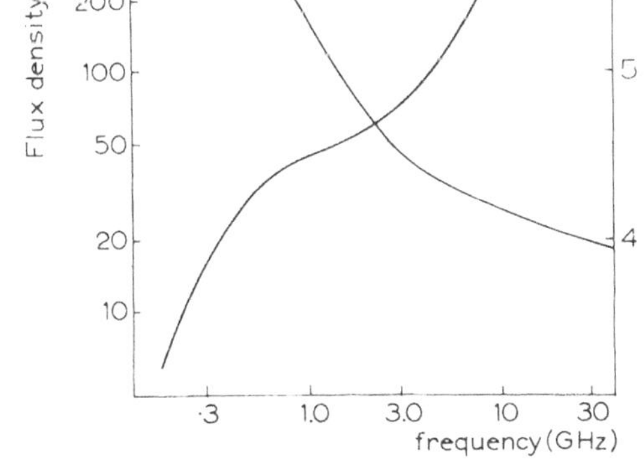

Fig. 11.1. Spectrum of the quiet Sun at sunspot minimum (after H. Tanaka *et al.*: 1973, *Solar Phys.* **29**, 243). Flux density F and equivalent brightness temperature are plotted.

Bruzek and Durrant (eds.), Illustrated Glossary for Solar and Solar-Terrestrial Physics. 111–138.

11.2. Slowly Varying Component or Sunspot Component (S-Component)

This is the enhanced thermal emission from active regions at centimeter and decimeter wavelengths. Its intensity changes relatively slowly and is strongly correlated with the number of sunspots present on the solar disk – hence 'sunspot component'. At large sunspot numbers the intensity is comparable to that of the quiet-Sun emission. On radioheliograms the sources of the slowly varying component – the radio plages – are a few minutes of arc in extent (depending on wavelength) with fine structures considerably smaller than $1'$ within them. They coincide with Ca plages and X-ray plages. The enhanced radio emission is due to enhanced electron density in the coronal condensation of the active region and to gyroresonance absorption. Consequently the radiation has a large contribution from coronal levels and a correspondingly high brightness temperature (see Figure 7.1c).

A relatively weak slowly varying component can be distinguished at shorter meter wavelengths on multi-element interferometer scans of the quiet Sun. This emission seems to stem from both helmet and active region coronal streamers. At 169 MHz the height of the source is $\approx 0.4\, R_\odot$.

Axisa, F., Avignon, Y., Martres, M. J., Pick, M., and Simon, P.: 1971, *Solar Phys.* **19**, 110.
Kundu, M. R.: 1965, *Solar Radio Astronomy*, John Wiley, New York, Ch. 6.

11.3. Microwave Burst

Microwave bursts occur at centimeter wavelengths often with extensions into the millimeter and decimeter domains. They are generally broadband with intensity peaks at different frequencies occurring roughly simultaneously. A wide variety of morphological forms exists.

A large proportion show a simple rise to a single peak and a subsequent decay. A *single burst* may be an impulsive burst with a very sudden onset and a steep rise in intensity, the peak being reached sometimes within one or a few seconds. The total duration is from one to a few minutes. Or a single burst may be long-enduring and exhibit a *gradual rise and fall.* A slow rise to a maximum intensity, generally less than 50 flux units, is followed by a gradual return to the original flux level. The total duration is $\approx$ 15–60 min.

Another common form shows two or more intensity peaks and is described as a *complex burst.* Sometimes an intense burst is preceded by a gradual increase in flux, a *precursor.* An enhanced level succeeding a burst is known as a *post-burst increase* (Figure 11.2).

A small fraction of microwave bursts reach flux densities in excess of 500 flux units. They generally show great complexity and are designated *great bursts.* Most represent the microwave part of a type IV burst.

Different classes of spectra at peak intensity can be distinguished: (1) steady increase of intensity with frequency, (2) steady decrease of intensity with frequency, (3) a spectral maximum, usually in the cm range, (4) a U-shaped spectrum (type U) with a minimum roughly between 500 and 2000 MHz, more typical of large, flare-associated (commonly type IV) events and indicative of a proton event.

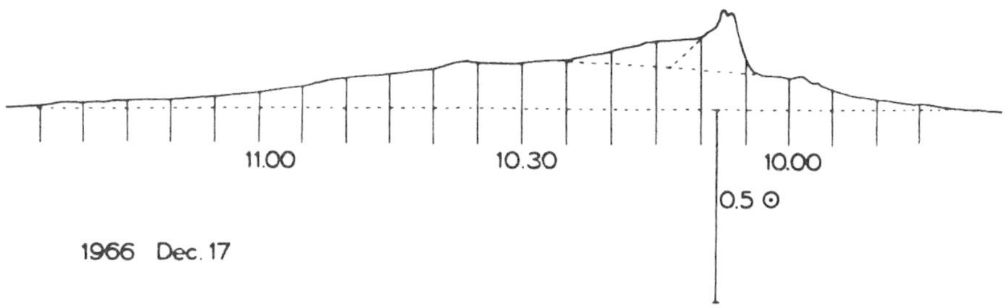

Fig. 11.2. A combination of microwave events on a record taken at 3000 MHz. The record shows: – a gradual rise and fall, lasting from 0943 to 1125 UT; – a slightly complex burst starting at 1004 UT; – a post-burst increase, ending at 1025 UT. The record was obtained at the Nera Observatory, the Netherlands.

The sources have dimensions $\approx 1'$ to $4'$ but show fine structures on a scale of $10''$. A bipolar polarization structure is not uncommon. With the probable exception of the long-enduring bursts of relatively small intensity the radiation is non-thermal in origin. It is mostly gyro-synchrotron radiation produced by barely to mildly relativistic electrons.

Almost invariably a microwave burst is associated with an optical and/or an X-ray flare.

Basu, D. and Covington, A. E.: 1968, *Solar Phys.* **5**, 102.
Castelli, J. P., Aarons, J., and Michael, G. A.: 1967, *J. Geophys. Res.* **72**, 5491.
Covington, A. E.: 1951, *J. Roy. Astron. Soc. Can.* **45**, 15.
Guidice, D. A. and Castelli, J. P.: 1973, in R. Ramaty and R. G. Stone (eds.), *Proc. Symp. on High Energy Phenomena on the Sun*, Goddard Space Flight Center, Greenbelt, p. 87.
Hachenberg, O.: 1965, in J. Aarons (ed.), *Solar System Radio Astronomy*, Plenum Press, New York, Ch. 12.
Kundu, M. R.: 1965, *Solar Radio Astronomy*, John Wiley, New York, Ch. 7.

11.4. Post-Burst Decrease

On rare occasions the intensity immediately after a microwave burst drops to a level slightly lower than the pre-burst level. In the course of several minutes the intensity returns to its original level. It has been suggested that it is due to absorption by surge-like material overlying the source of the slowly varying component of the microwave emission.

Recently, another type of *negative burst* associated with a moving dark filament has been detected.

Covington, A. E. and Dodson, H. W.: 1953, *J. Roy. Astron. Soc. Can.* **37**, 207.
Covington, A. E.: 1973, *Solar Phys.* **33**, 439.

11.5. Microwave Pulsations

A train of relatively small pulses sometimes appears superposed on the overall course of intensity variation in a microwave burst. The pulse intervals are ≈ 10–20 s and typically 5 pulses occur in a train. On a steep rise or fall of overall intensity such pulses may give the record the appearance of a staircase. Correlations with similar pulsations in an accompanying X-ray burst have been observed on a few occasions.

It has been suggested that the pulsations result from an electron cyclotron wave train bouncing on both ends of a magnetic loop.

Janssens, T. J., White, III, K. P., and Broussard, R. M.: 1973, *Solar Phys.* **31**, 207.
Maxwell, A. and Fitzwilliam, J.: 1973, *Astrophys, Letters* **13**, 237.

11.6. Noise Storms

Two classes of noise storms are observed: type I storms and decameter storms:

(a) The *Type I storm* is one of the commonest solar radio phenomena. Type I storm activity is usually concentrated in a frequency band of ≈ 100 MHz somewhere in the metric range (300–50 MHz). It comprises short bursts of small bandwidth (type I bursts) generally superposed on a background of slowly varying radiation (background continuum). The intensities vary enormously, from a small fraction of the quiet-Sun intensity up to several tens or even more than one hundred times that value. The relative contributions to the intensity of the bursts and the background also differ widely from one storm to another. The durations of noise storms generally range between several hours and a few days. The height of the source region is $\approx 0.3\ R_\odot$ at 200 MHz and increases with decreasing frequency probably in parallel with the change of plasma frequency with height. Most noise storms are strongly circularly polarised (close to 100%) but unpolarized storms occur sporadically, particularly if the storm is observed near the solar limb. They are invariably associated with sunspots, especially spots in the central part of the solar disk (see Figure 11.3).

(b) *Decameter storms* occur at frequencies below roughly 40 MHz. The background continuum is less developed than in most type I storms and the bursts are mainly decameter type III, type IIIb, split-pair, drifting-pair and fast-drift storm bursts. The storms have characteristics that vary with the position of the storm on the disk. At large and intermediate solar longitudes type IIIb bursts tend to dominate whilst near the central part of the disk type III bursts are relatively predominant. Drifting-pair bursts appear only at a narrow range of longitudes about the central meridian. This behaviour presumably reflects the different directivities of the various mechanisms. There is a rather strong correlation between the occurrences of type I and decameter storms. Decameter type III bursts sometimes seem to grow out of chains of type I bursts.

Fokker, A. D.: 1965, in J. Aarons (ed.), *Solar System Radio Astronomy*, Plenum Press, New York, Ch. 9.
Møller-Pedersen, B.: 1974, *Astron. Astrophys.* **37**, 163.

Fig. 11.3. Radio spectrogram of a noise storm, with the background continuum subtracted (May 5, 1971, 0717 UT). This record was obtained with the Dwingeloo 60-channel radio spectrograph. Time runs from left to right, seconds are marked.

11.7. Radio Continuum

(a) The *background continuum* in type I noise storms is the (variable) intensity level from which, on a single-frequency record, the storm bursts can be individually distinguished. It has been suggested in the early literature that the background represents the superposition of many indistinguishable small bursts, but this does not really seem to be the case. In strong storms the background continuum may reach (considerably) more than 10 times the intensity of the quiet-Sun radio emission.

(b) The *type IV continuum* is the broad-band continuum of type IV bursts. Microwave, decimetric and metric continua can be distinguished. The radiation is produced by the gyro-synchrotron mechanism and by the scattering of Čerenkov plasma waves on the background thermal plasma.

11.8. Type I Burst or Storm Burst

This is the class of short-lived, narrow-band bursts that usually occur in great numbers during a noise storm (Figure 11.3). They occur at meter wavelengths; at 200 MHz they last 0.3–0.7 s and have a bandwidth of 3–5 MHz. Their peak intensities vary widely, from a fraction of the quiet-Sun intensity to many times that value. They may appear "tilted" in the (f-t)-plane, with e.g. $\mathrm{d}(\ln f)/\mathrm{d}t \simeq \pm 0.05\ \mathrm{min}^{-1}$. Storm bursts are normally strongly circularly polarized but occasionally unpolarized ones are observed, predominantly near the solar limb. The angular sizes of storm bursts are 2–4′ at 200 MHz and 3–5′ at 80 MHz.

About 15% of type I bursts show a drift in position at rates of 1–3 arcmin s^{-1} (at 170 MHz). These *drifting-type I bursts* occur more frequently near the limb than near the centre of the solar disk.

Type I burst chains occur frequently during noise storms. They consist of from several up to hundreds of storm bursts arranged in relatively narrow-band (5–15 MHz) lanes that often drift slowly to (mostly) lower frequencies.

No satisfactory explanation for type I bursts is yet available.

Bougeret, J. L.: 1973, *Astron. Astrophys.* **24**, 53.
Elgarøy, Ø.: 1965, in J. Aarons (ed.), *Solar System Astronomy*, Plenum Press, New York, Ch. 10.
Elgarøy, Ø. and Ugland, O.: 1970, *Astron. Astrophys.* **5**, 372.
Hanasz, J.: 1966, *Australian J. Phys.* **19**, 635.

11.9. Type II Burst or Slow-Drift Burst

This is the class of large events at meter/decameter wavelengths that drift from high to low frequencies at a rate $\approx \frac{1}{4}\ \mathrm{MHz\ s}^{-1}$ (Figure 11.4). They occur in association with some 25% of the more important flares, starting about 10 min after the flash phase and lasting about 10 min. About half the cases show harmonic structure, a fundamental and second harmonic emission band (with relative bandwidth $\Delta f/f \approx 0.1$) showing a detailed resemblance in fine structure, including split bands. The harmonic ratio is generally slightly below 2.0 (average 1.95).

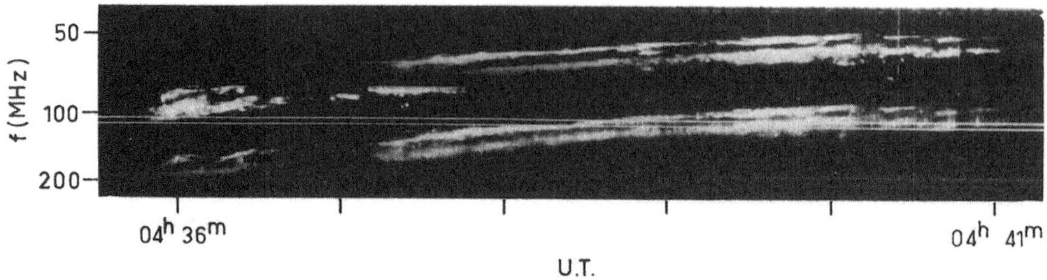

Fig. 11.4. Radio spectrogram of a type II event. Harmonic and split-band structure are very distinct in this example which was recorded with the Culgoora radio spectrograph. (After G. A. Dulk: 1970, *Proc. Astron. Soc. Australia* **1**, 308.)

They have been detected from a space vehicle at a frequency as low as 0.3 MHz corresponding to a distance of more than $30R_\odot$ from the Sun. On the plasma hypothesis, the drift rate corresponds to a speed $\leqslant 10^3$ km s^{-1}. The moving agency is considered to be a collisionless magnetohydrodynamic shock wave produced at the time of the flare flash phase. Such a shock is emitted over a wide angle, as appears from the large size of the type II burst source which may extend in a huge arc around the flare (Figure 11.5).

Emanating from a type II band in a dynamic spectrogram are sometimes fast-drifting, type III-like bursts with both forward and reverse drift. These give rise to a pattern known as *herringbone structure.*

The harmonic bands of a type II burst are sometimes split identically into two or more components separated by about 10% of their mid-frequency (*split bands*). It has been suggested that this structure results from emission by plasma both ahead of and behind a type II shock front (Figure 11.4).

Roberts, J. A.: 1959, *Australian J. Phys.* **12**, 327.
Smerd, S. F., Sheridan, K. V., and Stewart, R. T.: 1975, *Astrophys. Letters* **16**, 23.
Wild, J. P. and Smerd, S. F.: 1972, *Ann. Rev. Astron. Astrophys.* **10**, 159.

11.10. Fast-Drift Storm Burst

This is a category, distinguished by Ellis, of decameter bursts having an instantaneous bandwidth of only ≈ 0.03 MHz with a drift towards lower frequencies of 1–2 MHz s^{-1}. They occur at frequencies below 50 MHz during decameter noise storms, often in groups of 10 or so. Their total duration is 1–2 s.

Ellis, G. R. A.: 1969, *Australian J. Phys.* **22**, 177.

11.11. Drifting Pair

A drifting pair is a decameter (20–70 MHz) phenomenon which consists of two parts, the second being a repetition of the first after a delay of 1½–2 s. They show a frequency drift

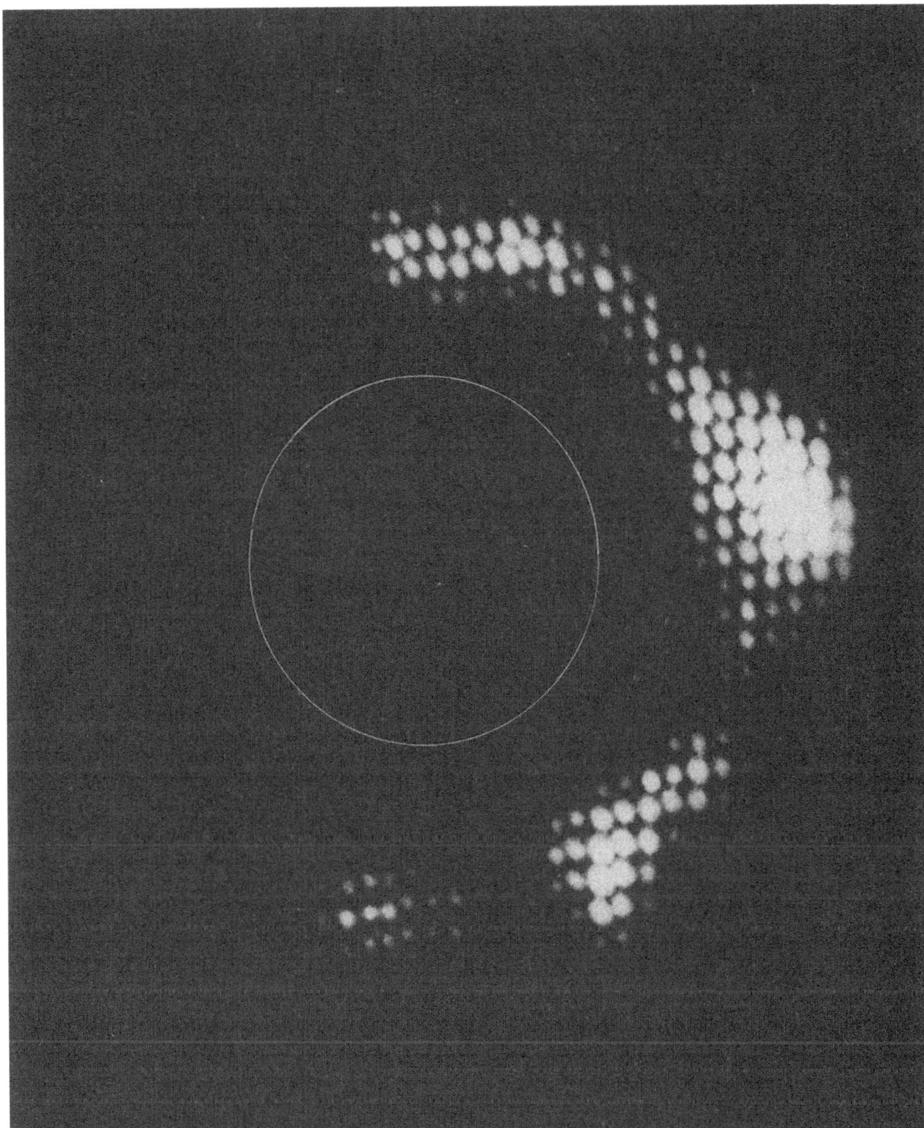

Fig. 11.5. Radio heliogram at 80 MHz of a type II burst source, obtained with the Culgoora radio heliograph on March 30, 1969. (After S. F. Smerd: 1970, *Proc. Astron. Soc. Australia* **1**, 305.)

rate ≈ 2 MHz s^{-1}. The drift may be towards lower frequencies (forward) or towards higher frequencies (*reverse-drifting pair*). The duration at any single frequency is usually less than 1 s.

A *hook burst* consists of a trace of forward drift with slope and bandwidth similar to that of a drifting pair, which suddenly reverses into a trace of increasing frequency.

Ellis, G. R. A.: 1969, *Australian J. Phys.* **22**, 177.
Roberts, J. A.: 1958, *Australian J. Phys.* **11**, 215.

11.12. Type III Burst or Fast-Drift Burst

This class of burst is a very common phenomenon at meter/decameter wavelengths. In a dynamic spectrogram they present a sharply defined front which drifts rapidly from high to low frequencies at a rate described by

$$\frac{df}{dt} = -0.01 f^{1.84}. \qquad (f \text{ in MHz s}^{-1})$$

Their durations are ≈ 1 or a few seconds, increasing with decreasing frequency. The peak intensities are generally high, of the order of 100 flux units or considerably more. They occur frequently in groups, the stronger ones often in huge complexes. At decameter wavelengths they can be so numerous as to constitute a storm. Type III bursts may show some circular polarization though most have none. Several authors claim to find linear polarization when measuring with small bandwidths.

Type III emission is ascribed to the scattering of Čerenkov plasma waves produced by fast electron streams with speeds $\geqslant \frac{1}{2}c$. It is believed that by far the most type III bursts are observed at the second harmonic of the local plasma frequency and that the first harmonic, if it exists at all, occurs only sporadically (Figure 11.6).

Lin, R. P. (ed.): 1976, *Solar Phys.* **46**, 433.
Stewart, R. T.: 1974, in G. Newkirk (ed.), 'Coronal Disturbances', *IAU Symp.* **57**, 161.
Various authors: 1974, *Space Sci. Rev.* **16**, (1/2); Special Issue on 'Type III Solar Radio Bursts and Solar Electrons in Space'.

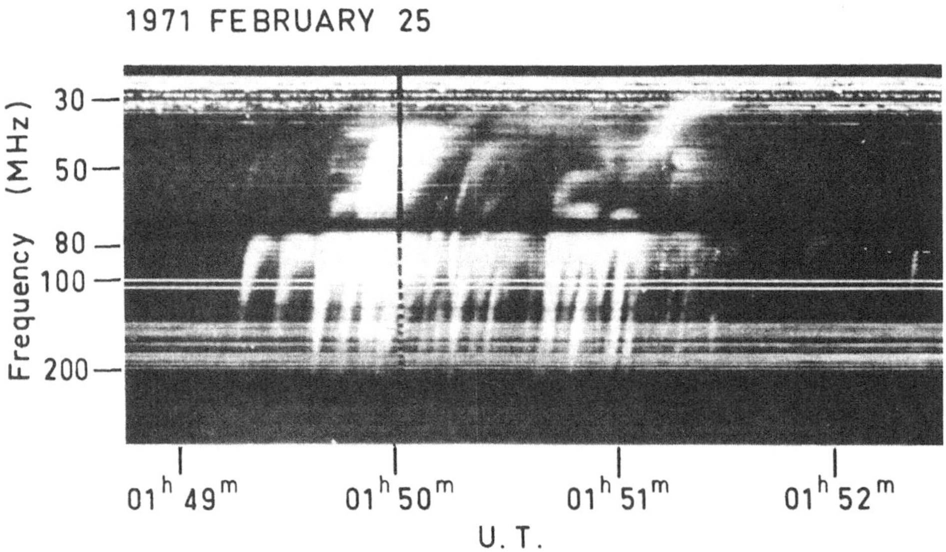

Fig. 11.6. A group of type III bursts, containing also some *J*-bursts. The record was obtained with the Culgoora radio spectrograph. (After I. D. Palmer and R. P. Lin: 1972, *Proc. Astron. Soc. Australia* **2**, 101.)

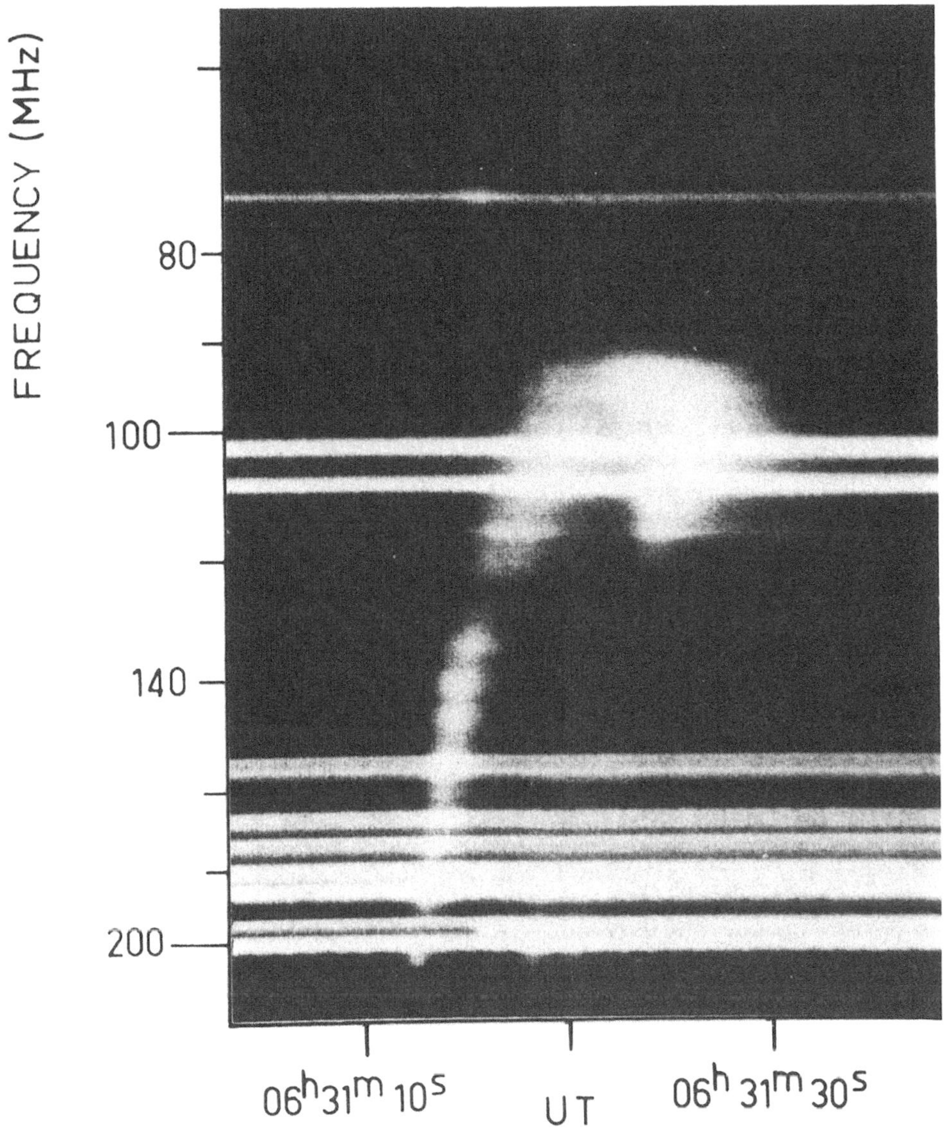

Fig. 11.7. Example of a U-burst, recorded with the Culgoora spectrograph. (After R. T. Stewart: 1975, *Solar Phys.* **40**, 417.)

11.13. U-Burst

This is a variant of the type III burst that appears as an inverted letter U on a dynamic spectrogram (Figure 11.7). The frequency drift rate at the high frequency starting point is comparable to that of the type III bursts. The turnover frequency may be anywhere in the meter/decameter domain. One as low as 1 MHz has been observed from a space vehicle. The duration is of the order of a few up to several seconds, being generally longer at lower turning frequencies. The progressive turning and reversal is due to a deflection of the exciting electrons towards denser coronal regions by a magnetic arch configuration.

Normally the return stroke is less intense than the forward. If the burst extends hardly further than its turn-over point it is called a *J-burst* (Figure 11.6).

Labrum, N. R. and Stewart, R. T.: 1970, *Proc. Astron. Soc. Australia* **1**, 316.
Stone, R. G. and Fainberg, J.: 1971, *Solar Phys.* **20**, 106.

11.14. Stria Burst

These are short (≈ 1 s) bursts with a narrow bandwidth (15–100 kHz) and often a slow frequency drift ($\leqslant 0.07$ MHz s^{-1}) towards lower frequencies. They occur in the frequency range <20–70 MHz. Stria bursts can occur singly, in split pairs (*split-pair bursts*) or in triplets (Figure 11.8). The frequency separation of the components in pairs or triplets lies in the range 40–300 kHz.

There also exists a class of stria bursts of longer duration (4–18 s) and no frequency drift known as *diffuse stria bursts.*

Sometimes two analogous stria bursts occur as a kind of echo event with a time delay of 1–4 s.

Chains of stria bursts having frequencies <20–70 MHz and drifting in a fashion similar to that of type III bursts have been named *type IIIb bursts* (Figure 11.9). They often appear as a kind of precursor to a normal type III burst separated from it by an interval of $\leqslant 10$ s. They have a duration of about one second rather than the 3–8 s length of the accompanying type III burst, and have a much higher degree of circular polarization, up to 100%.

Baselyan, L. L., Goncharov, N. Yu., Zaitsev, V. V., Zinichev, V. A., Rapoport, V. O., and Tsybko, Ya. G.: 1974, *Solar Phys.* **39**, 213.
de la Noë, J. and Boischot, A.: 1972, *Astron. Astrophys.* **20**, 55.
Ellis, G. R. A. and McCulloch, P. M.: 1967, *Australian J. Phys.* **20**, 583.

11.15. Type IV Burst

This is a prolonged flare-associated radio event that covers a broad band. Major type IV events cover the whole radio spectrum from centimeter (or even millimeter) wavelengths up to decameter wavelengths. The complete type IV event has components with different places of origin and different radiation mechanisms.

The *microwave type IV* (IVμ) is normally a microwave burst of the 'great burst' type. The mechanism is very probably (gyro-) synchrotron radiation; the sense of the (partial) polarization corresponds to that of the extraordinary mode.

The *decimeter type IV* (IVdm) extends from roughly 2000–200 MHz. Its sense of polarization seems to correspond to that of the ordinary mode and it is probably produced by scattering of Čerenkov plasma waves. The decimeter continuum often shows peculiar fine structures – absorptions, pulsating structure, etc. (Figure 11.10).

At meter wavelengths there are two types of sources, the *moving* (IVmB) and the

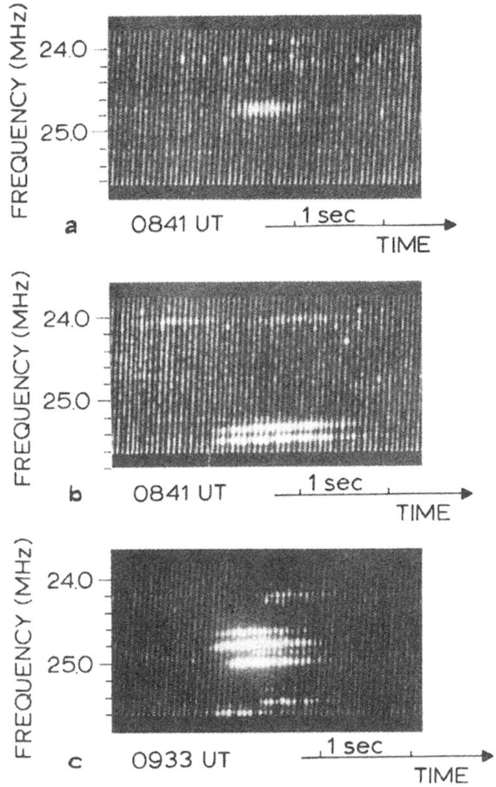

Fig. 11.8. Examples of stria bursts. (a) singlet, (b) split pair, (c) triplet. (After L. L. Baselyan *et al.*: 1974, *Solar Phys.* **39**, 213.)

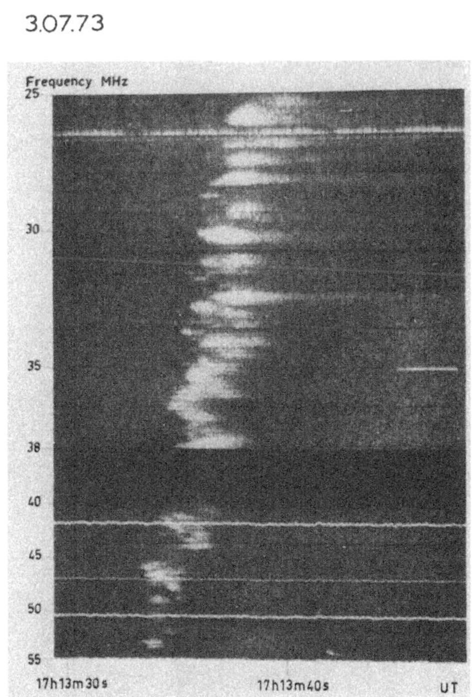

Fig. 11.9. Example of a type IIIb burst. This record was obtained from a radio-spectrograph connected with the 1000 ft Arecibo dish. (After J. de la Noë and A. Boischot: 1972, *Astron. Astrophys.* **20**, 55.)

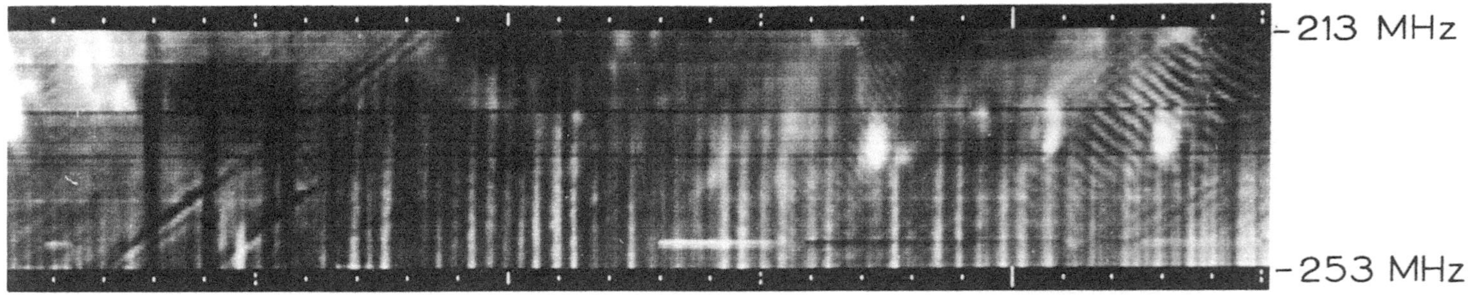

Fig. 11.10. Fine structures in the type IV continuum (event of June 29, 1971). One can distinguish: – broad-band short-lived absorptions (to the left); – pulsating structure (best seen left from centre); – zebra pattern or parallel drifting bands (to the right); – intermediate-drift bursts or fibres (best seen at left, crossing the absorptions). This record was obtained with the Dwingeloo 60-channel radiospectrograph with the background continuum subtracted. Time runs from left to right, seconds are marked.

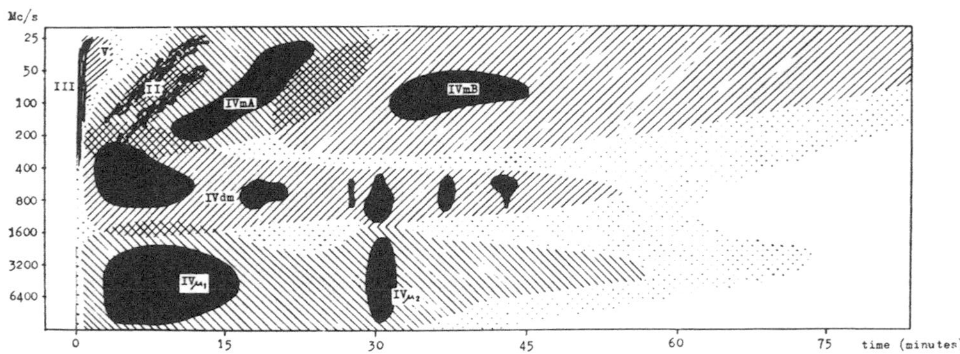

Fig. 11.11. Outline of the components of a type IV event (schematic spectral diagram).

stationary type (IVmA). Radiation from a moving source is seen not infrequently immediately after a type II burst. Stationary sources are situated at a relatively low level and may persist for several hours. Sometimes they develop gradually into a type I noise storm. An idealised picture of the arrangement of the components in the time-frequency plane is given in Figure 11.11.

Kundu, M. R.: 1965, *Solar Radio Astronomy*, John Wiley, New York, Ch. 11.

11.16. Moving Type IV Burst

Moving type IV bursts are continuum bursts at meter wavelengths produced by outward moving sources. They are usually associated with erupting prominences. Three types of moving sources have been distinguished with the Culgoora radioheliograph.

(a) Isolated sources which move out to several solar radii and probably correspond to a kind of self-contained plasmoid.

(b) Expanding arches which are apparently arranged along magnetic field structures tied into the photosphere (Figure 11.12); near their feet polarizations are opposite.

(c) Advancing fronts which are probably connected with a shock.

Source speeds are of the order of a few hundred km s^{-1}, but fronts may advance at about 1000 km s^{-1}.

The emission is generally ascribed to gyro-synchrotron radiation from electrons with energies $\approx 0.1-1$ MeV.

Smerd, S. F. and Dulk, G. A.: 1971, in R. Howard (ed.), 'Solar Magnetic Fields', *IAU Symp.* **43**, 616.
Weiss, A. A.: 1963, *Australian J. Phys.* **16**, 526.

11.17. Pulsating Structure

Pulsating structure is the name given to relatively broad-band (several tens to hundreds of MHz), quasi-periodic intensity fluctuations in the decimeter/meter continuum of type IV bursts. The fluctuations are generally almost in phase over their full bandwidth. The quasi period typically amounts to 0.5 to 2 s.

The modulation height is typically of the order of a few dB.

There is no strict division between pulsating structures and quasi-periodic sequences of broad-band absorptions (Figure 11.10).

Gotwols, B. L.: 1972, *Solar Phys.* **25**, 232.
Rosenberg, H.: 1970, *Astron. Astrophys.* **9**, 232.
Young, C. W., Spencer, C. L., Moreton, G. E., and Roberts, J. A.: 1961, *Astrophys. J.* **133**, 243.

11.18. Absorptions (Broad-Band, Short-Lived)

These are sudden depressions of intensity in the decimeter/meter continuum of type IV bursts that occur practically instantaneously over a broad frequency band of one octave, giving the impression that part of the continuum is suddenly switched off and on. The

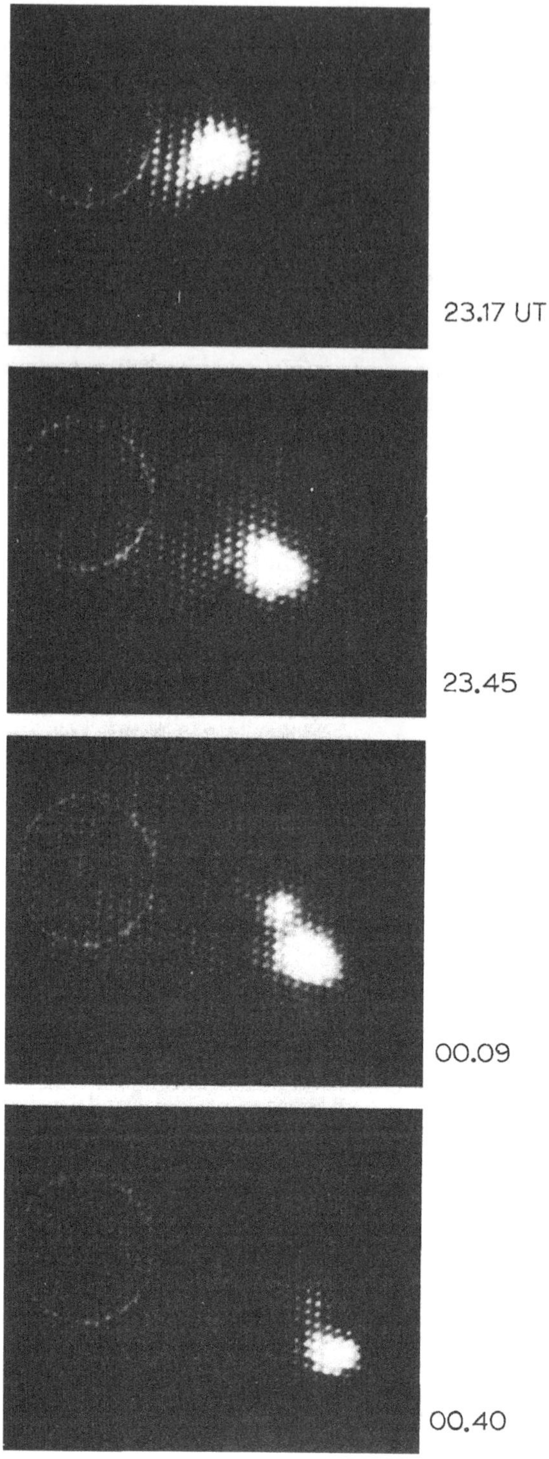

Fig. 11.12. Radioheliograms at 80 MHz of a moving type IV burst obtained with the Culgoora radio heliograph on March 1, 1969. (A. C. Riddle: 1970, *Solar Phys.* **13**, 448.)

depressions are sometimes deeper than −3 dB and the durations range between 0.1 and about 1 s. Broad-band absorptions often occur in sequences that may be either quasi-periodic (like pulsating structure) or irregular (Figure 11.10).

An explanation has been suggested in terms of a temporary interruption of the loss-cone instability due to the passage of fast particles parallel to the magnetic field.

Kuÿpers, J.: 1974, *Solar Phys.* **36**, 157.

11.19. Zebra Pattern or Parallel Drifting Bands

This feature is sometimes present in the decimeter/meter type IV continuum and consists of approximately equidistant parallel drifting emission ridges separated by troughs of depressed intensity. The separation of adjacent ridges is ≈ 2–4 MHz (Figure 11.10). It has been suggested that they are due to plasma waves at the upper hybrid frequency excited under certain circumstances at places where that frequency is an integral multiple of the gyrofrequency.

Elgarøy, Ø.: 1961, *Astrophys. Norv.* 7, 123.

11.20. Intermediate-Drift Bursts or Fibre Bursts

These bursts occur in type IV continua at decimeter/meter wavelengths as ridges formed by the adjacent depression and enhancement of intensity that drift in frequency at rates of −5 to −50 MHz s^{-1}, intermediate between those of type III and type II bursts. The enhancement is on the high frequency side of the ridge. The instantaneous bandwidth is ≈ 2 MHz and the range of frequencies covered is ≈ 50–150 MHz (see Figure 11.10).

It has been suggested that intermediate-drift bursts originate through the coupling of whistler waves, travelling along a magnetic field, with plasma waves.

Young, C. W., Spencer, C. L., Moreton, G. E. and Roberts, J. A.: 1961, *Astrophys. J.* **133**, 243.

11.21. Tadpoles

Slottje introduced this name to denote a very peculiar type of absorption-emission microstructure of the type IV continuum in the frequency range 220–320 MHz. It consists of a deep absorption 'body' with a bright emission 'eye' on the high frequency side and a relatively weak emission tail on the low frequency side covering a bandwidth ≈ 40 MHz. The structure is aligned roughly parallel to the frequency axis in a dynamic spectrogram. The lifetime of the 'body' is of the order of 0.2–0.3 s (Figure 11.13).

The 'eye' is perhaps the same as the *mini burst*, a type of narrow-band (1–3 MHz), short-lived (0.05–0.15 s) event observed at 500–550 MHz.

Elgarøy, Ø. and Sveen, O. P.: 1973, *Solar Phys.* **32**, 231.
Slottje, C.: 1972, *Solar Phys.* **25**, 210.

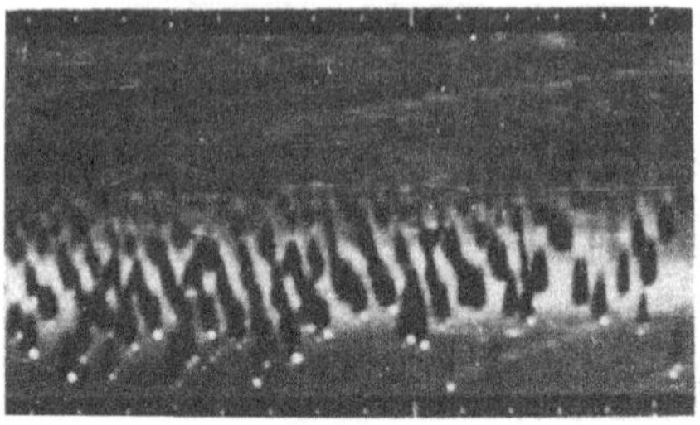

Fig. 11.13. Example of tadpoles. This record was obtained with the Dwingeloo 60-channel radiospectrograph, with the background continuum subtracted (March 2, 1970). Time runs from left to right, seconds are marked.

11.22. Spike Burst or Flash Burst

These are bursts at meter wavelengths with an extremely short duration ≈ 0.1 s or less. Roughly two categories exist: Storm-like spike bursts have bandwidths of a few up to about 10 MHz, not unlike type I bursts. They are concentrated in rather short time intervals of a few minutes duration and tend to occur in the later phases of type IV events.

Type III-related spike bursts occur just before and at slightly higher frequency than a type III burst so that it appears aligned with the type III burst in the frequency-time plane. They show a range of bandwidths from a few up to a few tens of MHz. Most show a frequency drift of more than 100 MHz s^{-1}.

Spike bursts are fully circularly polarized.

de Groot, T.: 1966, *Rech. Astron. Obs. Utrecht* **18**, 1.
Eckhoff, H. K.: 1966, *Inst. Theor. Astrophys.*, Oslo, Rep. No. 18.
Tarnstrom, G. L. and Philip, K. W.: 1972, *Astron. Astrophys.* **16**, 21; **17**, 267.

11.23. Type V Burst

This is the name for the broad-band continuum emission at the longer meter wavelengths that lasts for a minute or so after the occurrence of a type III burst. The sources of the type V bursts are observed to have a larger angular diameter than the sources of the concomitant type III and to be slightly displaced from them. It is assumed that the emission arises from Čerenkov plasma waves excited by electrons temporarily trapped in magnetic loops adjacent to the neutral plane in which the type III burst occurs (Figure 11.14).

Weiss, A. A. and Stewart, R. T.: 1965, *Australian J. Phys.* **18**, 143.

11.24. Distinctive Events

Reports of solar radio bursts observed at single frequencies are currently listed in the

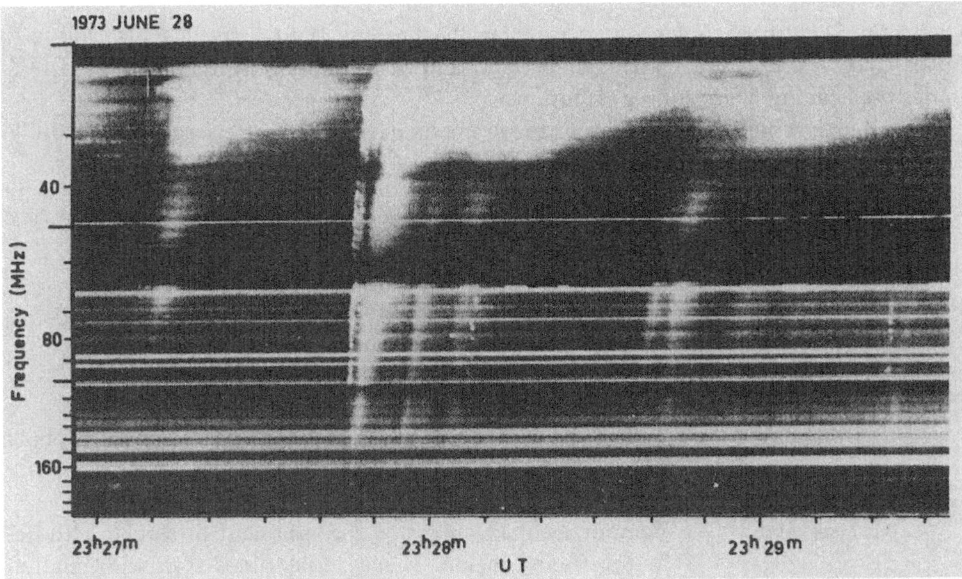

Fig. 11.14. Example of a combined type III-type V event, with a type IIIb precursor burst. This is a Culgoora radio spectrogram. (After T. Takakura and S. Yousef: 1975, *Solar Phys.* **40**, 421.)

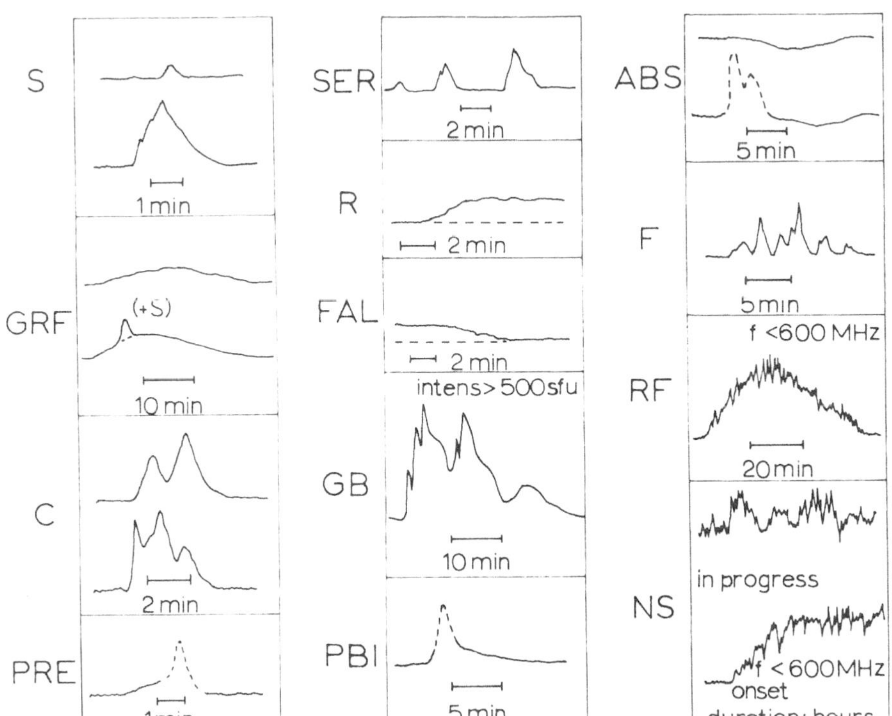

Fig. 11.15. Distinctive events and their designation.

Quarterly Bulletin on Solar Activity under the heading 'Distinctive Events' and in the *Solar Geophysical Data* (National Oceanic and Atmospheric Administration, U.S.A.) under the heading 'Outstanding Occurrences'.

The shapes which distinctive microwave bursts have on a single frequency record are designated by the following symbols (see Figure 11.15).

S = Simple:	Mostly non-thermal 'microwave impulsive burst' or 'decimetric burst'.
C = Complex:	Combination of a few or many simple bursts.
F = Fluctuation:	Minor C sometimes superposed on the main burst.
GB = Great Burst:	Major C of special importance.
PRE = Precursor:	Pre-burst activity connected to the main burst.
PBI = Post-Burst Increase:	Tail of the main burst which may be regarded as enhancement of S-component.
GRF = Gradual Rise and Fall:	Temporal enhancement of S-component or similar activation in the flaring region. It may sometimes start with relatively sharp rise like a simple burst. If this sharp rise can be clearly recognized as simple burst, GRF becomes PBI. Note that both have similar characteristics.
ABS = Absorption:	Absorption due to surge-like material mainly appears after the burst; it is sometimes called post-burst decrease. The same phenomenon may occur more frequently, but it can only be recognized when the flux comes down to pre-burst level. Temporal fall of flux, sometimes called negative burst, may be listed also as ABS, but it may simply be the temporal fall of emission.

The following three symbols are simply morphological, which may be necessary due to limited observing time or for the simplicity of tabulation:

R = Rise: This may also occur as the onset of long-enduring enhancement of S-component associated with other solar events.

FAL = Fall

SER = Series of bursts

On dm-m-Dm wavelengths most of the events are C, including F and GB. The following two symbols are used exclusively for this wavelength range:

NS = Noise Storm

RF = Rise and Fall: Defined as more or less irregular rise and fall of continuum with duration of the order of minutes to an hour.

In the *Solar-Geophysical Data* a number code is used in addition. This code is explained in the SGD explanation of data reports (= supplement to the February issue).

Instruction Manual for Monthly Report of Solar Radio Emission, World Data Center-C2 for Solar Radio Emission, Toyokawa Observatory (Japan), 1975.

11.25. Spectral Diagram

A spectral diagram is a contour representation of the development of a major radio event. The intensity is depicted in a scale of different shadings, as a function of time and frequency. It gives a better overall picture of the course of an event than a radio spectrogram. It is derived from a (great) number of single frequency records but is necessarily rough and does not show short-lived or narrow-band details (Figure 11.16).

Krüger, A.: 1972, *Physics of Solar Continuum Radio Bursts*, Akademie Verlag, Berlin, Appendix 2.

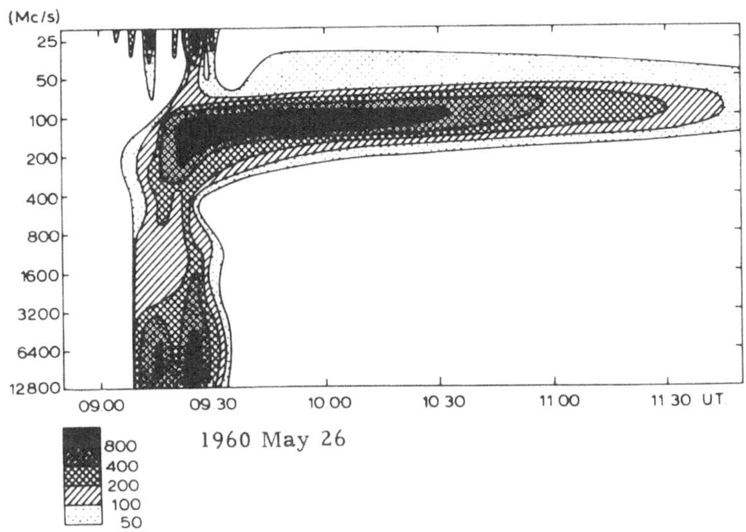

Fig. 11.16. Example of a spectral diagram that outlines a type IV event.

11.26. Directivity

The directivity of radiation from solar radio sources is the property of being beamed. Until the advent of the 'Stereo Experiment' the directivity could only be inferred statistically from studies of centre-to-limb variation of intensity and/or frequency of occurrence. For example,

(a) noise storms: They occur infrequently and with small intensities when associated with centres of activity further than 60° from the centre of the disk.

(b) microwave slowly-varying component: The centre-to-limb variation is somewhat broader than a cosine variation.

(c) type I bursts: The Stereo Experiment has revealed that the 3 dB full beamwidth is less than 25°.

(d) type III emission: This shows much less directivity.

The *Stereo Experiment* was the enterprise that detected and measured the directivity of solar burst radiation at 169 MHz by simultaneous observations from Nançay and the Soviet planetary probe Mars-3, during the period May 31, 1971 until February 24, 1972.

Caroubalos, C. and Steinberg, J. L.: 1974, *Astron. Astrophys.* **32**, 245.
Steinberg, J. L., Caroubalos, C., and Bougeret, J. L.: 1974, *Astron. Astrophys.* **37**, 109.

11.27. Solar Radio Telescopes

Most solar radio telescopes collect the radiation by one or more parabolic dishes. At the longer meter and the decameter wavelengths the use of arrays of dipole, yagi, helical or log-periodic antennas is not uncommon. In the focus of the dish the radio signal is commonly fed to a horn (cm/dm wavelengths), to a dipole or dipole-array or to a log-periodic antenna (dm/m wavelengths).

Sizes of dishes range from about 1 m to some 25 m. Radiospectrography is usually performed with a single dish. Radio heliography requires an interferometric combination of single antennas (dishes or other types, depending on wavelength). A variety of antenna configurations is found among heliographic radio telescopes. A few solar radio telescopes of different designs follow as examples.

8 cm Radioheliograph of Toyokawa Observatory (Res. Inst. of Atmospherics, Nagoya Univ., Japan)

This radioheliograph consists of a *T*-shaped array of 3 m parabolic dishes, with 32 elements in the E-W direction and 17 elements in the N-S direction. It has a maximum resolving power of 1'.5.

Two-dimensional scanning produces a map every 40 s. A computer system performs the data acquisition and on-line data processing.

Grating radio telescope of Algonquin Radio Observatory (Lake Traverse, Ont., Canada)

This radio telescope consists of an array of 32 three meter parabolic reflectors as one element of a compound interferometer and 4 similar reflectors placed symmetrically around the array along the same E-W line. The aperture of 665 m produces a fan beam of 0'.5 E-W by 2° N-S at the operating frequency of 2800 MHz.

The Nançay 169 MHz radioheliograph

This is an E-W array of two 10 m and sixteen 3 m dishes arranged as a compound interferometer along a baseline of 3200 m length. Correlations of both 10 m aerials with each of the 3 m aerials yield 32 separate Fourier components of the spatial brightness distribution. One-dimensional images with a resolution of 1'.7 are obtained by means of Fourier transformation at a rate of more than 10 s^{-1}.

The Culgoora radioheliograph

This radio telescope consists of an array of 96 dishes (diam. 13 m) placed along a circle with a diameter of 3 km. The beam of this array is modified by introducing in quick succession suitable delays in the antenna signals and combining the resulting signals.

A branching network provides for 48 outputs that correspond with a north-south row of 48 antenna beams. A system of computer controlled phase shifters performs the functions of scanning and tracking. At 80 MHz the Sun is pictured twice a second in a 2° x 1°.6 field comprising 60 x 48 points spaced by 2'.1.

The half power width of the pencil beam is 3'.7 at 80 MHz.

Operating at 80 MHz since 1968, the heliograph has been converted to operate also at 160 MHz and at 43.25 MHz.

The Clark Lake array (California)

Conical log spiral antennas are arranged in a *T*-configuration comprising 480 elements along a 3000 m E-W arm and 240 elements along a 1800 m N-S arm. The elements are fixed in a vertical direction; they have a frequency-independent beamwidth of 100°. Beam positioning is accomplished by adjustment of the relative phases between the elements. The instrument is capable of operation at any frequency between 15 and 130 MHz with corresponding beamwidths of 27′ to 3′. Pictures are obtained once every second.

Proc. IEEE **61** (9), 1973, Special Issue on 'Radio and Radar Astronomy'.

11.28. Radio Interferometer

A radio interferometer consists of two or more aerials that are interconnected by cables so that the signals received at each aerial interfere.

With a two-element interferometer one can determine

(a) the position of the 'centre of gravity' of the solar radio disk;

(b) the fringe visibility of the source.

Lobe-sweeping interferometers sweep the lobe pattern by inserting a continuous phase shifter in one cable (*phase-shifting interferometer*). This device has been applied to locate transient sources on the Sun.

A *multi-element* or *grating interferometer* consists of 2^n equidistant elements (where n is normally 3 to 5). The antenna pattern comprises a limited number of relatively narrow beams each of which can scan a source of limited extent, such as the solar disk. It is possible to use the elements of a grating interferometer for aperture synthesis. Fast synthesizing permits quasi-instantaneous mapping of the Sun as is performed at 169 MHz with the 32-element interferometer at Nançay.

A *compound interferometer* is a combination of two aerial systems connected to one another in the phase-switching mode. One system is commonly a grating; the other may be a grating, a two-element (unswitched) interferometer or a single dish. If arranged in a suitable configuration along a straight line it can yield a narrow beam in an economic way – without covering the whole baseline with equidistant antennas.

The combination, by a phase switching device, of two arrays of aerials at right angles to one another (East-West and North-South) is called a *Christiansen-cross* or *grating cross*. Its antenna pattern consists of a number of well-isolated pencil beams. Such interferometers have been extensively used to map the Sun at cm and dm wavelengths. A circular array of antennas is employed in the Culgoora radioheliograph.

Christiansen, W. N. and Högbom, J. A.: 1969, *Radiotelescopes*, Cambridge.
Covington, A. E., Legg, T. H., and Bell, M. B.: 1967, *Solar Phys.* **1**, 465.
Kraus, J. D.: 1966, *Radio Astronomy*, McGraw-Hill Book Co. New York, Ch. 6.

11.29. Radio Polarimeter

The polarization state of solar radio emission is measured by a polarimeter. In solar phenomena circular polarization is predominant so most solar polarimeters are designed to measure the sense and degree of circular polarization. Radiation is received with two orthogonally crossed dipoles or by dual linearly polarized horns in the focus of a parabolic antenna. By introducing a phase delay of $\pm 90°$ between the signals and summing them one obtains either the right-handed or left-handed circular component. By phase-switching and synchronous detection it is possible to measure the difference of intensities of the two components with great precision.

The measurement of the general state of polarization requires the determination of the polarization ellipse (axial ratio and orientation) or, equivalently, the four Stokes parameters. This requires in addition to the two orthogonal linear receptors, a third in a diagonal direction. Four independent measurements then suffice to specify the polarization state fully.

Akabane, K. and Cohen, M. H.: 1961, *Astrophys. J.* **133**, 258.
Grognard, R. J. M. and McLean, D. J.: 1973, *Solar Phys.* **29**, 149.
Tanaka, H. and Kakinuma, T.: 1959, in R. N. Bracewell (ed.), 'Radio Astronomy', *IAU Symp.* **9**, 215.

11.30. Radiospectrograph

This instrument records the instantaneous spectral intensity distribution of solar radio events as a function of time. Radiospectrography usually employs the *swept-frequency technique* in which the receiving frequency is swept over a certain band by the rapid tuning of the local oscillator. The signal intensity is used to modulate the brightness of the light spot of a cathode ray tube. By imaging the CRT screen on a moving film a time-frequency display of the dynamic development of the spectrum is obtained. The film record is called a *dynamic spectrogram.* The first radiospectrograph of Wild and McCready covered the frequency range 70–130 MHz. A drawback of this technique is the restricted sensitivity as the radiation at any one frequency is received for only a small fraction of the time.

Multi-channel spectrographs overcome this by having a collection of receivers tuned to a sequence of nearby frequencies but suffer from the difficulty of producing identical receiver characteristics over a large amplitude range. The contrast of relatively weak signals can be considerably increased by subtracting the quasi-steady background intensity. The limited dynamic range (20–30 dB) allowed by the use of photographic film for display can be extended by employing digital magnetic tape recording which has the additional advantage of yielding accurate quantitative measurements.

Cole has introduced a quite different technique which uses a transducer bonded to a fused silica rod to convert the wide-band (100 MHz) radio signal into a travelling acoustic wave. The pressure variations of this wave modulate the phase of a collimated laser beam passing transversely through the rod. The diffracted light in the emergent beam, when imaged by a lens, is the required spectrum.

Cole, T. W.: 1973, *Proc. Inst. Elec. Electron. Engrs.* **61**, 1321.

Cole, T. W.: 1973, *Astrophys. Letters* **15**, 59.
de Groot, T. and van Nieuwkoop, J.: 1968, *Solar Phys.* **4**, 332.
Wild, J. P. and McCready, L. L.: 1950, *Australian J. Sci. Res.* A3, 387.

11.31 Čerenkov Emission

Čerenkov emission is produced if a charged particle moves in a medium at a speed greater than the local phase speed of some wave mode. The wavefronts form the Čerenkov emission cone. The particle energy may be lost to either transverse (propagation vector perpendicular to the electric vector) electromagnetic waves (Čerenkov radiation) or to longitudinal (propagation parallel to the electric vector) electrostatic waves (*Čerenkov plasma waves*).

In the solar corona the phase speed of plasma waves is less than the speed of light so that a beam of fast electrons ($v \gtrsim \frac{1}{3}c$), as invoked to explain type III bursts, would excite Čerenkov plasma waves along its path.

11.32 Landau Damping

The longitudinal electric fields of plasma waves will interact with those plasma electrons having speeds close to the wave phase speed. This is a collisionless, collective effect. If the speeds of such resonant electrons are less than the wave speed the wave loses energy to the particles; if they are greater, the electrons give up energy to the wave by the Čerenkov process. Thus if slower resonant electrons are the more numerous the wave will be damped. In a thermal plasma all plasma waves are Landau damped. The damping increases exponentially as the wave speed approaches the thermal speed of the electrons and so sets an upper limit to the frequency spectrum of the plasma waves.

If the faster resonant electrons are more numerous, which requires the electron velocity distribution to have more than one maximum, there can be *coherent wave growth* at the expense of the fast electron stream. This is a two-stream instability.

Exactly analogous interactions take place between the rotating electric fields of cyclotron waves and the gyrating electrons producing them and result in *cyclotron damping* in thermal plasmas.

Stix, T. H.: 1962, *The Theory of Plasma Waves*, McGraw-Hill Book Co., New York.

11.33. Gyroresonance Absorption

Since electrons emit gyro-synchrotron radiation at frequencies $f = sf_H$ ($s = 1, 2, 3, \ldots$, f_H = gyrofrequency), they will also act as absorbers of incident radiation at the gyrofrequency and its harmonics. This is gyroresonance absorption which plays an important role for the slowly varying component and for the microwave bursts.

In those coronal layers in which the magnetic field strength is such that $f = sf_H$ ($s = 1, 2, 3, \ldots$) the optical thickness to incident radiation is one or two orders of magnitude greater for the extraordinary component than for the ordinary, and decreases rapidly with increasing s. These are the reasons why (a) the observed extraordinary component of the slowly varying radiation originates above the level where $s = 4$, whereas the ordinary component can reach us from the $s = 3$ level, and (b) the radiation of microwave bursts is received at frequencies at least 3 or 4 times the value of the gyrofrequency in the source region.

Zheleznyakov, V. V.: 1970, *Radio Emission of the Sun and Planets*, Pergamon Press, Oxford, Sec. 26, 29.

11.34 Gyro-Synchrotron Radiation

A non-relativistic electron moving in the presence of a magnetic field, B, gyrates about the field direction at the gyrofrequency f_H [MHz] = 2.8 B [G]. A 'slow' electron will then radiate only at the gyrofrequency in the em mode polarized in the sense of rotation of the electron (the extraordinary component). This mode cannot escape from the solar atmosphere. Fast electrons radiate also at harmonics of the gyrofrequency in both the extraordinary and ordinary (polarization counter to the sense of electron gyration) components. The higher the speed the more harmonics are emitted. For electron energies $E < mc^2$ the radiation is called *gyroradiation* and is restricted to relatively low order harmonics. For energies $E > mc^2$ the emission extends over a large number of harmonics and is then known as *synchrotron radiation.* Individual harmonics from highly relativistic electrons in an assembly are smeared out into a continuum.

The radiation observed from (moving) type IV bursts and from microwave bursts is most probably of an intermediate type, from mildly relativistic electrons, called gyro-synchrotron radiation. It is produced at magnetic field strengths of the order of a few gauss (type IV) or of a few to several hundred gauss (microwave).

Takakura, T.: 1960, *Publ. Astron. Soc. Japan* **12**, 352.
Wild, J. P., Smerd, S. F., and Weiss, A. A.: 1963, *Ann. Rev. Astron. Astrophys.* **1**, 291.

11.35. Faraday Rotation

In a plasma in the presence of a magnetic field a polarized transverse electromagnetic wave propagates in two independent modes. The modes have different phase velocities and opposite circular polarization at high frequencies. Thus the plane of polarization of a linearly polarized wave suffers rotation on passing through the medium given by

$$\Delta\phi \text{ [radians]} = 2.36 \times 10^4 f^{-2} N B_{\|} \Delta L,$$

where

f ≡ frequency [Hz],
N ≡ electron density [cm^{-3}],
$B_{\|}$ ≡ component of magnetic field parallel to the direction of propagation [G],
ΔL ≡ path length [cm].

Whenever linearly polarized radiation suffers Faraday rotation, the frequency dependence of the rotation produces a spread of polarization position angles within a given bandwidth. Hence the measured degree of polarization is less than the intrinsic degree

(*Faraday depolarization*). The amount of Faraday rotation can be inferred by comparing the degree of polarization measured in two different bandwidths.

Attempts have been made to trace the effects of depolarization on type III bursts.

Akabane, K. and Cohen, M. H.: 1961, *Astrophys. J.* **133**, 258.

11.36. Loss Cone

An axially-symmetric magnetic field configuration with two constrictions (a '*magnetic bottle*') is capable of trapping charged particles by reflecting their motions in the regions of increased field strength. However, at any point, those particles whose velocity vector lies within a certain solid angle about the field direction will penetrate the mirror and be lost from the bottle. This solid angle forms the loss cone which has a semi-angle, α, given by

$$\sin^2 \alpha = B/B_m,$$

where

B $\equiv$ the local magnetic field strength;

B_m $\equiv$ the maximum magnetic field strength of the bottle.

Collisions will continually scatter particles into the loss cone and so drain the bottle.

If fast electrons moving through a background plasma are trapped in a magnetic bottle a loss cone is established and the electron velocity distribution becomes anisotropic. This situation is unstable and waves tending to restore the equilibrium are excited. A variety of waves may be produced including plasma waves near the upper hybrid frequency moving preferentially across the field. This is one of a class of instabilities due to anisotropy of the velocity distribution function known as *loss-cone* or *Harris instabilities.* They may play a role in the excitation of the Type IV dm continuum and in the production of some of its fine structure.

Kuijpers, J.: 1974, *Solar Phys.* **36**, 157.

11.37. Mode Coupling

Coupling between the two em modes in a plasma if frequently referred to in discussions of the polarization of solar radio phenomena. Modes may couple if a radio wave passes through a region in which the magnetic field is nearly at right angles to the direction of propagation, a quasi-transverse region. A mode preserves its identity if its frequency is less than the transition frequency, f_t [MHz], given by

$$f_t^4 = 5\, L_H f_H^3 f_p^2,$$

where

f_p $\equiv$ the plasma frequency [MHz];

f_H $\equiv$ the gyrofrequency [MHz];

L_H $\equiv$ the magnetic scale length [km], $L_H^{-1} = \nabla\theta$;

θ $\equiv$ the angle between the wave normal and the magnetic field direction.

In this case the sense of rotation of the wave reverses if the direction of the magnetic field with respect to the direction of propagation reverses. If the frequency is greater than f_t the wave continues as if the magnetic field were not there and preserves its sense of rotation in spite of any field reversal.

Cohen, M. H.: 1960, *Astrophys. J.* **131**, 664.
Zheleznyakov, V. V.: 1970, *Radio Emission of the Sun and Planets*, Pergamon Press, Ch. 6.

11.38. Plasma Hypothesis

This is the assumption that the radio emission of certain bursts is due to the conversion by scattering of energy from longitudinal plasma waves into transverse electromagnetic radiation at a frequency slightly above the local plasma frequency. It permits an estimate, given a coronal model, of the velocity of the exciting agent from the frequency drift rate of the fundamental of a type II burst.

11.39. Plasma Wave or Langmuir Wave

Plasma waves are longitudinal electrostatic waves produced by oscillations of electrons in a warm (collisional) plasma. Their frequency, f, is related to the wavenumber, k, by

$$f^2 = f_p^2 + \left(\frac{kV_e}{2\pi}\right)^2,$$

where

f_p ≡ the plasma frequency, f_p [MHz] $= 8.9 \times 10^{-3} N_e^{1/2}$;
N_e ≡ the electron density [cm^{-3}];
V_e ≡ the electron thermal speed, V_e [cm s^{-1}] $= 6.7 \times 10^5 T_e^{1/2}$;
T_e ≡ the electron temperature [K].

In cold plasmas ($V_e = 0$) the electric field is the self-consistent and coherent field due to the electrons behaving collectively. Non-propagating oscillations at the plasma frequency follow. At high temperatures the collective behaviour is disrupted and an acoustic-type wave results.

In the presence of a magnetic field the gyrations of the electrons modify the waves propagating at an angle, θ, to the field. Then the dispersion relation is

$$f^2 = f_p^2 + f_H^2 \sin^2 \theta + \left(\frac{kV_e}{2\pi}\right)^2 \quad \text{for} \quad \frac{2\pi f}{k} < c,$$

where f_H ≡ the gyrofrequency.

At right angles to the field the wave has a pseudo-resonance at the *upper hybrid frequency*, f_{uh}

$$f_{uh}^2 = f_p^2 + f_H^2.$$

11.40. Quasi-Linear Relaxation or Plateau Formation

This process is frequently referred to in theories of the excitation of type III bursts by electron streams. Resonant plasma waves are excited by the two-stream instability which, in turn, produce random accelerations and retardations of the electrons causing them to diffuse in velocity space. This smoothes the electron velocity distribution function gradient which was responsible for the instability and damps it.

Davidson, R. C.: 1972, *Methods in Nonlinear Plasma Theory*, Academic Press, Ch. 9.
Kaplan, S. A. and Tsytovich, V. N.: 1967, *Astron. Zh.* **44**, 1194.
Kaplan, S. A. and Tsytovich, V. N.: 1968, *Soviet Astron.* **11**, 956.

11.41. Razin Effect

This is the suppression of (gyro-)synchrotron emission at the low-frequency side of the synchrotron spectrum due to the influence of the ambient plasma. The suppression becomes effective at frequencies for which $(1 - n^2)\gamma^2 \gtrsim 1$, where n is the refractive index and $\gamma = E/mc^2$ the Lorentz factor.

The spectrum produced by mildly relativistic electrons is determined by γ and the parameter $\alpha = 3\, f_{\mathrm{H}}/2\, f_p$. If $\alpha\gamma < 1$ the emission is strongly suppressed at low frequencies but if $\alpha\gamma > 1$, it is relatively unaffected. For a given γ the cut-off shifts to higher frequencies as α decreases.

Equivalently, the spectrum will be significantly affected if the critical frequency f_c,

$$f_c\,[\mathrm{MHz}] = 20\, N_e/B_\perp,$$

where N_e is the electron density [cm^{-3}], and $B_\perp$ the component of magnetic field (G) perpendicular to the line of sight, is sufficiently large to lie within the (vacuum) radiation's main spectral range.

The Razin effect has been invoked to account for the low-frequency spectral profile of certain *metric type IV* and *microwave bursts.*

Ginzburg, V. L. and Syrovatskii, S. I.: 1965, *Ann. Rev. Astron. Astrophys.* **3**, 297.
Ramaty, R.: 1968, *J. Geophys. Res.* **73**, 3573.

11.42. Scattering (Radio)

In the context of solar radio astronomy the term 'scattering' is used with two different meanings.

(a) Scattering is the irregular refraction of radiation from solar radio sources by inhomogeneities in the distribution of electron density. It gives a point-like source an apparent extent which may be of the order of 1′ depending on wavelength and height of emission. It also produces the time dilatation (of a small fraction of a second) of an instantaneous burst that is emitted as an impulse of negligible duration.

The process was discovered from occultations of the Taurus A (Crab nebula) radio source by the corona.

(b) Scattering is also the process by which longitudinal plasma waves are converted into transverse electromagnetic radiation by interaction with neutral or charge density fluctuations. Fluctuations are present in any thermal plasma and the usual picture is of plasma waves being scattered by the so-called polarization clouds (constituting the thermal ion-acoustic turbulence) surrounding individual ions. The frequency difference between the two modes is $\sim k\ V_i$, where $k \equiv$ the wave number of the plasma wave, and $V_i \equiv$ the ion thermal speed.

Thus the radio emission produced has a frequency close to the plasma frequency.

Waves can also scatter on one another (coalesce) producing a single wave whose frequency and wave vector are the sums of those of the incident waves. This process producing radio emission at approximately twice the plasma frequency is called *combination scattering.*

Hewish, A.: 1955, *Proc. Roy. Soc.* **A228**, 238.
Riddle, A. C.: 1974, *Solar Phys.* **35**, 153.
Steinberg, J. L., Aubier-Giraud, M., Leblanc, Y., and Boischot, A.: 1971, *Astron. Astrophys.* **10**, 362.
Tsytovich, V. N.: 1970, *Nonlinear Effects in Plasmas*, Plenum Press.
Zheleznyakov, V. V.: 1970, *Radio Emission of the Sun and Planets*, Pergamon Press.

12. GENERAL THEORETICAL TERMS

C. J. DURRANT

12.1. Anomalous Dispersion

At frequencies close to that of a spectral line, a gaseous medium shows not only a rapid variation with frequency of absorption coefficient but also of refractive index. This anomalous dispersion is particularly important in the analysis of lines formed in the presence of a magnetic field. The differently split components of the line have different dispersions which leads to a rotation of the axes of the polarization ellipse as the radiation traverses the medium.

Stenflo, J. O.: 1971, in R. Howard (ed.), 'Solar Magnetic Fields', *IAU Symp.* **43**, 108.

12.2. Contrast Mechanisms

All solar spectral lines vary both in time and from point to point on the surface. The fractional variation of the line profile with respect to some reference profile is known as the *line contrast.* The mechanisms that produce this contrast vary from line to line and may be used as a diagnostic tool.

In general, photospheric lines respond to local variations of both density and temperature. For chromospheric lines only the collisional processes reflect strongly local atmospheric variations. Photo-ionization is governed by the radiation field of the underlying (distant) photosphere. Thus collisionally controlled lines, e.g. Ca II K, are sensitive to local conditions whilst photo-ionization controlled lines, e.g. Hα, are only weakly so. Mottles and other small-scale structures in the chromosphere can have very different appearances in these two lines.

A further complication may be introduced if the structure is optically thin. Then the source function is determined essentially by the external radiation field in all cases.

Gebbie, K. B. and Steinitz, R.: 1974, in R. G. Athay (ed.), 'Chromospheric Fine Structure', *IAU Symp.* **56**, 55.

12.3. Dynamical Stability

A dynamical system is unstable if any perturbation of the system causes a motion of growing amplitude. If such motion is aperiodic the system is said to be convectively unstable, if oscillatory to be overstable. When all perturbations are damped the system is stable but only marginally stable if any motion continues without growth or decay. Usually only infinitesimal perturbations are considered but some systems can be put permanently into a new configuration by only a finite amplitude perturbation. This occurs should the system occupy a local minimum of potential energy rather than an absolute minimum. Such systems are known as metastable.

Bruzek and Durrant (eds.), Illustrated Glossary for Solar and Solar-Terrestrial Physics. 139–147.

It should be stressed that stability analyses describe only the onset of instability. They do not describe any subsequent steady-state configuration.

Ledoux, P. and Walraven, T.: 1958, *Handbuch der Physik* **51**, 353.

12.4. Hanle Effect or Level Crossing Interference

When the separation of energy sublevels is less than their width they become degenerate states. Then the scattering of radiation involving these levels cannot be regarded simply as an emission following an absorption. There is a phase relationship between the incoming and outgoing radiation producing coherence in frequency, direction and polarization.

The presence of a magnetic field is just one factor leading to this effect. A magnetic field causes most magnetic sublevels to shift relative to one another which alters the coherence properties. In particular, the degree and direction of linear polarization is a function of the magnetic field strength and can be used to estimate the latter.

House, L. L.: 1971, in R. Howard (ed.), 'Solar Magnetic Fields', *IAU Symp.* **43**, 130.

12.5. Lighthill Mechanism

Lighthill estimated the production of pressure (acoustic) waves by fully developed, homogeneous and isotropic turbulence in a compressible medium. It is not demonstrably applicable to the solar convection zone but has nevertheless been applied. The resulting fluxes are very sensitive functions of the structure of the convection zone which is poorly known and so urges further caution.

Generally speaking, sound production is very inefficient because of the mismatch between the convective speeds and the sound speed.

Stein, R. F.: 1968, *Astrophys. J.* **154**, 297.

12.6. Line Control

Although the scattering term of a line source function is nearly always numerically the largest, it is the source and sink terms that differentiate the line formation mechanisms. They are said to control the line. Resonance lines in the solar atmosphere may be roughly classified according to whether

(i) the direct source and sink terms are much greater than the indirect; such lines, e.g. of ionized metals, are collisionally controlled,

(ii) the direct terms are much less than the photo-ionization (indirect) terms; such lines, e.g. of neutral metals, are photo-ionization controlled,

(iii) other cases, e.g. H, He lines, are mixed.

The scheme can be extended to subsidiary lines if they are formed in regions where the resonance transitions are optically thick (radiative rates in detailed balance) and where their collisional rates may be ignored. The H Balmer lines are such a case.

A few lines, such as intercombination lines, have very small radiative transition probabilities. Then the direct collisional source and sink terms may be dominant and the source function reduces to the Planck function determined by the local temperature. Such lines are said to be formed in LTE but one should note that the emission and absorption coefficients (and hence the optical depth scale) do not separately have their LTE values.

Jefferies, J. T.: 1968, *Spectral Line Formation*, Blaisdell, Waltham.

12.7. Macro/Microturbulence

The observable effect of motions, whether random or organized, in the solar atmosphere is governed by their scale-size. If the scale of variation of the velocity is much greater than unit optical thickness in a spectral line the whole column of gas emitting the line radiation moves coherently and the line will show only an appropriate Doppler shift. If the scale is much less, the motions act like an increased thermal motion and produce an extra broadening and hence strengthening of medium and strong lines but no net line shift. The former is known as macroturbulence and the latter as microturbulence. Of course, if the horizontal scale of the macroturbulence is less than the spatial resolution of an observation again no net line shift will be seen, just broadening without strengthening. This is called unresolved macroturbulence.

In fact, motions in the Sun occur on all scales so there are marked velocity gradients in all line forming regions. Then neither simple picture is appropriate and reliable quantitative estimates of the mechanical energy density and flux cannot be derived from them. Intermediate scale velocity fields, *mesovelocities*, can yield very different effects depending on whether lines are formed in LTE or NLTE.

Shine, R. A.: 1975, *Astrophys. J.* **202**, 543.

12.8. Magnetic Buoyancy

A horizontal magnetic flux tube below the solar photosphere is subject to a buoyant force. The gas pressures exterior and interior to the flux tube are related by

$$P_{ex} = P_{in} + H^2/8\pi,$$

where H is the magnetic field strength. Similarly

$$\rho_{ex} = \rho_{in} + \frac{\mu}{kT}\frac{H^2}{8\pi}$$

so that the density of the flux tube matter is lower than that outside, and an upward buoyant force results. This force is opposed by tension along the field lines.

Parker, E. N.: 1955, *Astrophys. J.* **121**, 491.

12.9. Non-Thermal Velocities

Observations of the profiles of optically thin solar EUV emission lines and visible region chromospheric lines show that the line widths are in excess of those expected from broadening by thermal motions at the local electron temperature. Since the ion-electron collisional relaxation times in the chromosphere through to the inner corona, are rapid enough for T_e to equal T_{ion} the extra component is attributed to non-thermal motions. It is not yet known whether these motions represent the passage of ordered waves through the atmosphere or disordered motion that could be referred to as turbulence.

Boland, B. C., Dyer, E. P., Firth, J. G., Gabriel, A. H., Jones, B. B., Jordan, C., McWhirter, R. W. P., Monk, P., and Turner, R. F. 1975, *Monthly Notices Roy. Astron. Soc.* **171**, 697.

12.10. Overstability

Overstability occurs when a perturbation of a system can oscillate with growing amplitude. It requires two forces, one destabilising and the other stabilising, to act out of phase.

The classic example occurs in stars with an outer ionization zone (not necessarily convective). When compressed the base of the zone ionizes further. This absorbs heat (γ *mechanism*) and can reduce the opacity (κ *mechanism*). More heat flows into this layer than out of it; the compression does not relax and the layer acts as a driving region. At the top of the zone a compression reduces the ionization; the opposite processes occur so the layer acts as a damping region. If the latter region is far enough from the centre of the star for non-adiabatic effects to be apparent, i.e. the luminosity does not change immediately but is frozen-in to some extent, the two regions will not cancel one another but can reinforce the motion leading to a growth of the oscillation.

Another example is found in superadiabatic regions where the radiative relaxation time is very short – as occurs at the very top of the solar convection zone. The destabilising buoyancy force can then dissipate leaving unbalanced pressure or magnetic restoring forces. Oscillations will then build up if the acoustic or MHD waves produced by this instability are trapped within some zone of the Sun.

Cox, J. P.: 1967, in R. N. Thomas (ed.), 'Aerodynamic Phenomena in Stellar Atmospheres', *IAU Symp.* **28**, 3.
Moore, D. W. and Spiegel, E. A.: 1966, *Astrophys. J.* **143**, 871.

12.11. Penetrative Convection or Overshoot

If a superadiabatic region is overlaid by a subadiabatic one – as is the solar convection zone by the photosphere – the motions resulting from the instability are not confined to the lower layer but extend into the upper region as penetrative convection.

Two models of such motions in the photosphere have been proposed. One pictures them as counter cells produced by gravity waves excited by temperature gradients due to convection cells in the lower region. The other employs the meteorologists' model of thermals. Random blobs of convecting material with excess buoyancy and momentum will overshoot the boundary of the superadiabatic region. They are then dissipated by thermalizing, mixing and generation of gravity waves.

A hydrodynamic theory of penetration in compressible fluids is under development.

Experiments suggest that the highest fluctuating velocities develop in the layers immediately above the limit of the superadiabatic layer just where the granulation is observed in the solar atmosphere.

Graham, E.: 1977, in J. P. Zahn (ed.), 'Problems of Stellar Convection', *IAU Colloq.* **38**, in press.
Moore, D. W.: 1967, in R. N. Thomas (ed.), 'Aerodynamic Phenomena in Stellar Atmospheres', *IAU Symp.* **28**, 405.

12.12. Rotational Braking

Angular momentum is lost from the Sun when solar wind particles become decoupled from the co-rotating solar magnetic field. Up till that point the ionized particles are accelerated by the field increasing their angular momentum. The accelerating force is transmitted by the magnetic field lines back to the subphotospheric layers into which the field is tied by the high conductivity and large scale, and slowly brakes these layers. Since they lie in the convective zone, convective mixing, which carries angular momentum, ensures that the loss is spread over the whole zone. Whether the internal radiative core of the Sun need also be significantly slowed down is a matter of debate.

12.13. Source Function

The source function for any radiative process is the ratio of the volume emission coefficient to the linear absorption coefficient. It appears as the radiative energy source term in the transfer equation which relates the radiative intensity to optical depth.

An important example is the source function of a resonance line transition. Absorbed line photons produce an excited atomic state which may de-excite either by re-emitting a line photon (scattering), by direct electron collision or by an indirect process involving another energy level, often photo-ionization followed by recombination to the ground state. The last two processes remove line photons and are thus said to contribute *sink terms* to the denominator of the source function. New line photons result not only from scattering but also from radiative de-excitation of the upper level following direct electron collisional excitation or indirect excitation, again often recombination following photo-ionization from the ground state. These processes generate line photons and are said to contribute *source terms* to the numerator of the source function.

Thomas, R. N.: 1965, *Some Aspects of Non-Equilibrium Thermodynamics in the Presence of a Radiation Field*, Univ. of Colorado Press, Boulder.

12.14. Thermodynamic Equilibrium

A system in thermodynamic equilibrium (TE) has its energy equipartitioned between all its degrees of freedom. The various distribution functions are governed by a single thermodynamic quantity, the temperature, through the Planck and Saha-Boltzmann laws.

If there is a net energy flux, as in stars, TE cannot be fulfilled because some particles must be distinguished in order to carry the flow information. However if most atoms and photons undergo many interactions before encountering a part of the system where the state parameters are significantly different, we may consider infinitesimal parts of the system to be effectively isolated and hence equipartitioned at the local electron kinetic temperature. This is *local TE* (*LTE*).

When any species fails in this criterion, it couples parts of the system with quite different state parameters. Then, in general, only the electrons can be taken to be in equilibrium with a Maxwellian velocity distribution. The other distribution functions take no simple form. This is the case of *non-local TE* (*NLTE*) of which LTE and TE are special cases.

In TE every process which depopulates an atomic energy level is exactly cancelled by its inverse, a situation known as detailed balance. Most astrophysical systems, whilst not in TE, are usually in a quasi-steady state. Then the sum of all processes depopulating an energy level is balanced by the sum of their inverses. This is *statistical equilibrium* (*SE*).

Thomas, R. N.: 1965, *Some Aspects of Non-Equilibrium Thermodynamics in the Presence of a Radiation Field*, Univ. of Colorado Press, Boulder.

12.15. Turbulence

In astrophysical contexts this term is used rather loosely to describe any non-steady velocity fields. As the Reynolds number of a flow (the ratio of sizes of the typical pressure and viscous forces) increases its laminar motion becomes unstable and breaks up into smaller-scale motions. Initially these lead to repeatable time-dependent (unsteady laminar) flows but at higher Reynolds number they become more complex, less repeatable and less dependent on the initial conditions. This is turbulent flow and must be treated statistically. If the velocity pattern is totally irregular in space and time the turbulence is said to be fully developed. Its properties are difficult to determine except in very special cases such as isotropic, homogeneous and incompressible turbulence. Even then, further assumptions about the cascading of energy from large- to small-scale motions are needed to derive the eddy energy spectra of Kolmogorov or Heisenberg.

Turbulence might be expected in the solar atmosphere because of the very low molecular viscosity but it is clear that it is not describable by any simple special case.

Bradshaw, P.: 1976, in P. Bradshaw (ed.), *Turbulence*, Springer-Verlag, Berlin, Ch. 1.

12.16. Waves

When a fluid system in a steady state is given a small displacement a restoring force due to pressure, buoyancy, magnetic field, etc., may be created. The linear equations describing the resulting motions can then be combined to give a wave equation which has constant coefficients if the properties of the medium are sufficiently uniform. The resulting motions are independent plane waves whose spatial and angular frequencies k, ω, are related by a dispersion relation. In non-dissipative systems energy is conserved in the direction of propagation but some waves can have a varying energy density in other directions. They are exponentially damped in space and thus are called *evanescent waves.* In dissipative systems the energy of the waves may either grow or decay in time.

In k-ω space the dispersion relation defines certain loci which delineate wave eigenmodes for that system. An example identifying the modes for typical low coronal conditions is shown in Figure 12.1.

Often only a component of k, such as the horizontal, k_x, may be directly measured. Then it is convenient to use the dispersion relation to define planes in k_x-ω space where allowed modes lie. Figure 12.2 shows such a diagram for an isothermal ($T \approx 5000$ K), exponential, non-magnetic atmosphere. Above the upper curve, *acoustic* (p) *waves* in which pressure is the dominant restoring force occur. None are allowed below the *acoustic cut-off frequency*

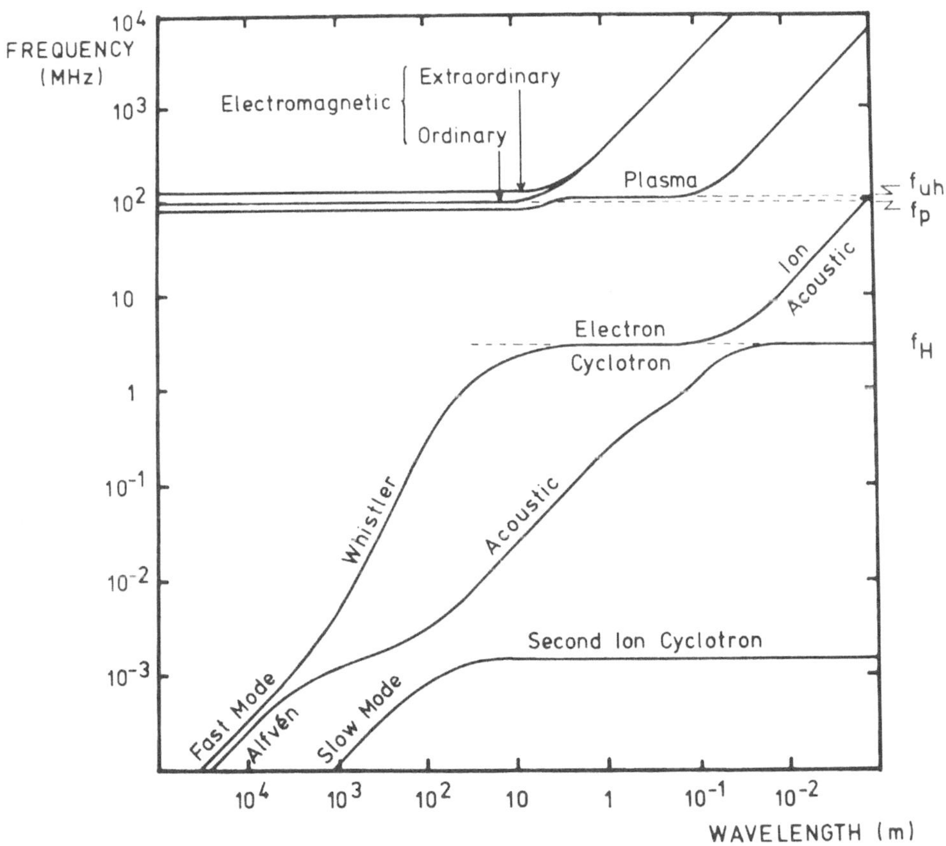

Fig. 12.1. Diagnostic diagram for a uniform isothermal medium with uniform magnetic field. The parameters have been taken to represent the low corona. $N_e = 10^8$ cm^{-3}; $T = 10^6$ K; $B = 1$ G. The wave modes propagating at an angle of $\pi/4$ to the magnetic field are identified.

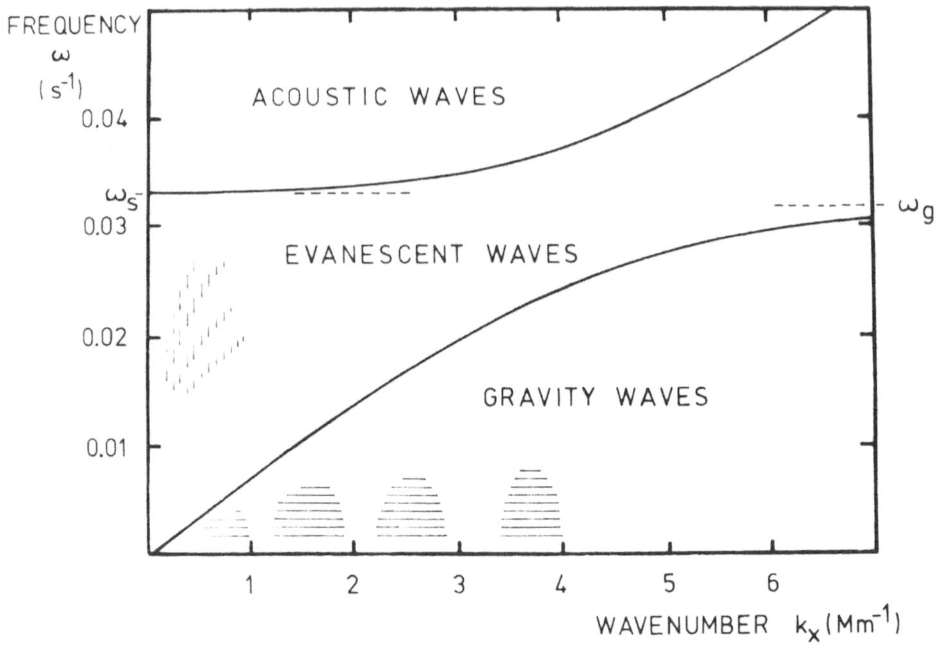

Fig. 12.2. Diagnostic diagram for an isothermal (5000 K), exponential atmosphere showing the location of the acoustic, gravity and evanescent modes. Shading denotes regions where velocity power is observed in the solar atmosphere – vertical, the 300 s oscillatory modes and horizontal, the granular and supergranular 'convective' components.

$$\omega_s = \frac{g\gamma}{2c_s},$$

where

g ≡ the acceleration due to gravity,

γ ≡ the ratio of specific heats,

c_s ≡ the sound speed.

Below the lower curve we have *internal gravity* (*g*) *waves* in which buoyancy is the major restoring force. None appear above the *Brunt-Väisälä frequency*

$$\omega_g = \frac{g}{c_s}(\gamma - 1)^{1/2}.$$

Elsewhere only evanescent waves exist.

Figures 12.1 and 12.2 are both examples of *diagnostic diagrams.*

Stein, R. F. and Leibacher, J.: 1970, *Ann. Rev. Astron. Astrophys.* **12**, 407.

12.17. Wave Tunnelling

As the properties of the solar atmosphere vary with height the boundaries of the acoustic and internal gravity wave regions in the diagnostic diagram shift. Thus for given k_x and ω there may be propagation allowed in the low photosphere and corona but only evanescent waves in some intermediate region. Such photospheric waves are trapped. However energy can leak across the barrier being carried by a combination of exponentially growing and decaying evanescent modes in a manner analogous to that of quantum-mechanical barrier penetration. This process is known as wave tunnelling.

Zhugzhda, Y. D.: 1972, *Solar Phys.* **25**, 329.

13. SOLAR WIND AND INTERPLANETARY MEDIUM*

L. SVALGAARD

13.1. Solar Wind

Due to high temperature and more importantly a small temperature gradient with height the solar corona is not in hydrostatic equilibrium but is continuously expanding into interplanetary space. The resulting outflow of coronal plasma is known as the solar wind. Because of the very high electrical conductivity of the coronal plasma the magnetic field in the corona controls both the heat conduction and the basic outflow of material. In regions where the magnetic field lines reach high into the corona (e.g. over coronal holes) the coronal expansion is greatly enhanced.

The expansion speed is very low in the inner corona but increases rapidly with height. At the *critical radius* the thermal energy and the expansion kinetic energy become comparable; at this point the velocity is close to the sound velocity in the plasma and the critical point is sometimes loosely referred to as the *sonic point.* At larger distances the expansion velocity increases still further and the solar wind becomes supersonic.

The solar magnetic field is now carried along with the expanding plasma resulting in an *interplanetary magnetic field.* At about $20 R_\odot$ distance from the Sun, the coronal expansion becomes very nearly radial but the rotation of the Sun winds the interplanetary magnetic field lines into Archimedes spirals on cones described by the rotating radius vector.

In early work on solar wind theory, incorrect assumptions about the electric field produced by polarization of the coronal plasma (Pannekoek-Rosseland electric field derived under the assumption of static equilibrium) led to evaporative models of coronal expansion predicting much lower (subsonic) velocities, i.e. a *solar breeze* rather than a solar wind. In situ measurements by spacecraft have confirmed the basic validity of the solar wind concept. Table 13.1 gives typical values of fundamental physical properties of the solar wind as measured near the Earth.

TABLE 13.1
Properties of solar wind

	minimum	average	maximum
Flux (10^8 ions cm^{-2} s^{-1})	1.0	3.0	100
Velocity (km s^{-1})	200	400	900
Density (ions cm^{-3})	0.4	6.5	100
Electron temperature (1000 K)	5	200	1000
Proton temperature (1000 K)	3	50	1000
Magnetic field strength ($\gamma = 10^{-5}$ G)	0.2	6.0	80
Alfvén speed (km s^{-1})	30	60	150
Helium abundance (fraction per number)	0.0	0.05	0.25

* This work was supported by the National Aeronautics and Space Administration under Contract NAS5-22993.

Bruzek and Durrant (eds.), Illustrated Glossary for Solar and Solar-Terrestrial Physics.
149–157.

Hundhausen, A. J.: *Coronal Expansion and the Solar Wind*, Springer Verlag, New York.
Parker, E. N.: 1958, *Astrophys. J.* **128**, 664.
Pneuman, G. W.: 1976, *J. Geophys. Res.* **81**, 5049.
Sonett, C. P., Coleman Jr., P. J., and Wilcox, J. M. (eds.): 1972, *Solar Wind*, NASA SP-308, Washington.

13.2. Solar Wind Models

Solar wind models may be classified as *fluid models* or *exospheric models* depending on the degree of coupling assumed between individual particles. The mean free path of solar wind protons in interplanetary space is of the order of 1 AU so that a collisionless treatment seems appropriate. On the other hand, the interplanetary magnetic field provides a strong coupling on a much smaller scale than 1 AU, justifying the use of a fluid description. Early exospheric models were unable to produce a supersonic solar wind; later refinements overcame this difficulty mainly by introducing a particle energy-dependent *critical level* or *exospheric base* (defining the level from which the particle can escape by evaporation) and by introducing regions of closed magnetic field lines that 'shut off' the escape of particles from certain coronal regions. A fundamental problem with these *evaporative models* is that they predict quite different evaporation (or expansion) speeds for ions with different charge-to-mass ratios. The observed near-equality of proton and α-particle mean velocities in the solar wind indicates that these ions are strongly coupled by some interaction mechanism and thus are not collisionless.

The first fluid models of the solar wind assumed that the plasma could be adequately described at each point by one density, one velocity and one temperature. Although such *one-fluid models* illustrate the basic phenomena in coronal expansion, they are only valid if the protons and the electrons interact strongly enough to maintain equipartition of thermal energy. Deep in the corona, coulomb collisions occur frequently enough to ensure such equipartition. As the plasma moves away from the Sun, collisions become so rare that equal temperatures for protons and electrons cannot be maintained. In fact, at 1 AU the electron temperature is about four times higher than the proton temperature (Table 13.1).

In *two-fluid models* the protons and the electrons are thermally decoupled. The electrons are strongly heated by thermal conduction throughout the flow, while the protons are unheated except for small contributions from the (much lower) proton thermal conduction and collisional energy exchange with electrons. As a result, the protons cool nearly as rapidly as in an adiabatic expansion. A basic problem with these cool protons is that they impede the coronal expansion so that extra heating of the corona is required far out into the solar wind rather than just near the Sun in order to reproduce the observed expansion rate. This *external heating* could be due to a combination of several effects. Dissipation of waves of various kinds has been considered. It is also possible that some non-collisional mechanism might exchange thermal energy between electrons and protons. The heating problem is one of the most debated in modern solar wind theory.

Asbridge, J. R., Bame, S. J., Feldman, W. C., and Montgomery, M. D.: 1976, *J. Geophys. Res.* **81**, 2719.
Barnes, A.: 1974, *Adv. Electronics Electron Phys.* **36**, 1.

Brandt, J. C. and Casinelli, J. P.: 1966, *Icarus* **5**, 47.
Hartle, R. E. and Sturrock, P. A.: 1968, *Astrophys. J.* **151**, 1155.
Holzer, T. E.: 1977, *J. Geophys. Res.* **82**, 23.
Jockers, K.: 1970, *Astron. Astrophys.* **6**, 219.

13.3. Angular Momentum Loss

The average flow direction of the interplanetary plasma is not quite radially away from the Sun. Rather, the solar wind appears to flow from a direction about 1.5° east of the Sun on the average. At low speeds (300 km s^{-1}) the flow appears to come even farther from the east (3°) shifting westward with increasing flow speed.

Any real non-radial component of the solar wind velocity implies a transport of angular momentum from the Sun to the interplanetary medium. The observed *azimuthal velocity component* corresponds to a torque on the Sun opposing solar rotation. This torque exerted on the Sun by the expanding coronal matter is sufficient to brake solar rotation in a time comparable to the age of the Sun ('magnetic braking').

Hundhausen, A. J., Bame, S. A., Asbridge, J. R., and Sydoriak, S. J.: 1970, *J. Geophys. Res.* **75**, 4643.
Weber, E. J. and Davis, L.: 1967, *Astrophys. J.* **148**, 217.

13.4. High-Speed Streams

A structureless coronal expansion (such as described by most models) is rarely observed. Rather, the solar wind exhibits large variations on a time-scale of several days, comparable to the basic time-scale of the overall expansion. The fundamental entity seems to be the high-speed stream in the sense that the solar wind as observed near the Earth often can be considered as a succession of such streams.

In a stream the flow speed rises from low (less than 400 km s^{-1}) values to a maximum (up to 800 km s^{-1}) in about two days. The subsequent decrease is often more gradual, although streams maintaining a high speed for several days are sometimes observed, especially during the declining phase of the sunspot cycle. The proton temperature varies in a way similar to the flow speed, while the electron temperature seems to show little or no variation within the streams. The density and the magnetic field strength rise to unusually high values near the leading edge of the stream. These peaks are generally followed by unusually low values (rarefaction region) persisting for several days within the stream. It is important to emphasize that this description of stream morphology is only valid near 1 AU from the Sun.

High-speed streams seem to flow out of regions in the corona where the magnetic field configuration is diverging and open. Such regions are often marked by the appearance of coronal holes. In addition, it follows that they are basically unipolar or at least of a dominant magnetic polarity. Such magnetic conditions are usually quite stable and persist for several solar rotations, giving rise to a pronounced tendency for high-speed streams to recur.

The motion of solar wind plasma streams with differing expansion speeds are strongly affected by solar rotation. A high-speed stream, distorted into a spiral by solar rotation will overtake and collide with any slower moving ambient solar wind material. The high electrical conductivity of the magnetized plasma prevents interpenetration of different plasma regimes with the result that the ambient or slower-moving plasma must be compressed and deflected to flow parallel to the interface with the fast stream. Similarly, the fast plasma will be slowed down and compressed somewhat at its leading edge. As the interface between the two plasma regions – the *interaction region* – propagates outward through the solar system, a system of shocks develops bounding the region: a *forward shock* travelling outward with respect to the interface and an associated *reverse shock* travelling inward. A large fraction of the mass associated with the stream is confined to the narrow shell bounded by the forward-reverse shock pair. Such shells are expected to be strong obstructions to the propagation of galactic cosmic rays.

It is significant that it is the presence of slower moving plasma that gives rise to the plasma compressions. As slow-moving solar wind tends to come from coronal regions with a more closed and confining magnetic configuration such as is associated with sunspots and general solar activity, we might expect more interaction regions and hence fewer cosmic rays when solar activity is high than when activity is low. It is probably important that the latitudinal extent of regions with closed field lines is at a maximum near solar maximum when the coronal holes over the polar regions disappear.

There seems to be a growing acceptance of the notion that a high-speed solar wind is the normal state of affairs and that slow-moving solar wind is a result of constricting forces in the corona preventing the free flow of the coronal expansion. The polar regions of the Sun which, during most of the sunspot cycle, are covered with unipolar coronal holes with diverging, open magnetic field lines should then be strong sources of very long-lived high-speed solar wind jets. Some support for this comes from radio observations of solar wind irregularities indicating a general increase of solar wind speed with latitude.

Baruch, E. and Sari, J. W.: 1976, *J. Geophys. Res.* **81**, 1453.
Burlaga, L. F.: 1975, *Space Sci. Rev.* **17**, 327.
Gosling, J. T., Hundhausen, A. J. and Bame, S. J.: 1976, *J. Geophys. Res.* **81**, 2111.
Hundhausen, A. J. and Gosling, J. T.: 1976, *J. Geophys. Res.* **81**, 1453.
Neugebauer, M. and Snyder, C. W.: 1966, *J. Geophys. Res.* **71**, 4469.

13.5. Mass Ejections

In addition to the steady coronal expansion, *transient ejections* of coronal material are often observed. One class of ejections is produced by solar flares. As the material moves rapidly outwards into the slower ambient solar wind, the plasma is compressed and pushed aside by the expanding flare ejecta. Generally a shock front will form at the leading edge of the compressed plasma shell. The compression region behind the shock is 0.1 to 0.2 AU thick. The tangential discontinuity that separates the compressed ambient solar wind from the flare ejecta is often followed by a thin (0.01–0.1 AU thick) *helium-rich* shell. Finally, a localized stream of high-speed, low-density solar wind often follows the ejection, lasting for a day or two. Although the mass and energy in a typical large ejection exceeds the mass and energy loss due to the normal 'quiet' solar wind expansion, the ejections occur so infrequently that they generally account for less than 10% of the total mass and energy transport in the solar wind.

Another class of mass ejections is related to eruptive prominences or disappearing filaments. About ten times less mass is involved but the ejections occur about ten times more often than flare-produced ejections. Plasma density enhancements (and associated magnetic field enhancements) due to both classes of ejections are strong scatterers of cosmic rays contributing to the modulation of galactic cosmic rays by solar activity (*Forbush decreases* of cosmic ray intensity).

Gosling, J. T., Hildner, E., Mac Queen, R. M., Munroe, R. H., Poland A. I., and Ross, C. L.: 1974, *J. Geophys. Res.* **79**, 4581.
Hirshberg, J., Bame, S. J., and Robbins, D. E.: 1971, *Solar Phys.* **18**, 313.
Hundhausen, A. J.: 1972 in Sonett, C. P., Coleman, P. J., and Wilcox, J. M. (eds.), *Solar Wind*, NASA SP-308, 393.

13.6. Waves in the Solar Wind

Observations indicate the essentially continuous presence of magnetic field fluctuations in the solar wind plasma that seem to be predominantly Alfvén waves being convected outward from the Sun. In addition to being convected the waves move outward from the Sun with respect to the plasma. Because of the outward propagation it must be presumed that these waves originate near the Sun – at least inside the critical radius. It has been suggested that the photospheric supergranulation is the source of these waves.

Belcher, J. W. and Solodyna, C. V.: 1975, *J. Geophys. Res.* **80**, 181.
Hollweg, J. V.: 1975, *Rev. Geophys. Space Phys.* **13**, 263.
Völk, H. J.: 1975, *Space Sci. Rev.* **17**, 255.

13.7. Discontinuities in the Solar Wind

Boundaries between distinct plasma regimes are convected along with the bulk motion of the solar wind. The plasma states on the two sides of such a boundary are not independent but are related by Maxwell's equations and the requirement for pressure balance normal to the boundary. For a *tangential discontinuity* the magnetic field and the flow velocity must be parallel to the boundary, but the magnitude and orientation of the two vectors can change in any manner consistent with this restriction. The tangential discontinuity is stationary with respect to the plasma.

Rotational discontinuities are sharp 'kinks' propagating along magnetic field lines at the Alfvén speed and both the magnetic field strength and the plasma velocity are the same on both sides of the discontinuity. Both types of discontinuities are very common features of the solar wind.

Filaments in the solar wind is a rather loose term implying that regions of the solar wind plasma are bounded by similarly oriented tangential discontinuities extending back to the Sun. This has been referred to as the *'spaghetti' model*. Observations seem to indicate that such filaments are rare and that the orientation of discontinuity surfaces generally changes from one surface to the next, giving the medium a discontinuous rather than filamentary structure.

Shocks form when two adjacent plasma regimes collide with relative speed exceeding the local sound (or Alfvén) speed. A rather complex set of possible shock types exists. An often occurring type is the *'fast shock'* in which the normal velocity component of the inflowing and outflowing material are respectively greater than and less than the fast MHD wave speed $V_F = (V_S^2 + V_A^2)^{1/2}$ where V_S and V_A are the sound and Alfvén speeds, but in which both normal velocity components are greater than the Alfvén speeds in the pre- and post-shock plasmas.

Two limiting classes of shock waves have been considered: *'driven waves'*, produced by prolonged and extended disturbances near the Sun, are characterized by post-shock increases in density, speed and temperatures of the driving gas behind the shock. The other type, *'blast waves'*, produced by impulsive disturbances near the Sun, are characterized by post-shock decreases in the density, speed and temperatures. Driven waves are associated with high-speed streams or with flares that release mass and energy into the solar wind for a time longer than about five hours, while interplanetary blast waves seem to be associated with flares that release energy for a time shorter than half an hour. Intermediate types of shock waves result from ejections with intermediate time scales.

Burlaga, L. F.: 1971, *J. Geophys. Res.* **76**, 4360.
Dryer, M.: 1975, *Space Sci. Rev.* **17**, 277.

13.8. Heliosphere

The solar wind extends into a region around the Sun appropriately called the *heliosphere* and is expected to terminate – at the *heliopause* – at a distance R at which the *ram pressure* of the wind balances the pressure in interstellar space. This interstellar pressure is due to a combination of contributions from the galactic magnetic field, cosmic rays, and interstellar gas. The value of R is not known but R = 50 to 100 AU is often quoted.

It is likely that the solar wind terminates with a shock transition from supersonic to subsonic expansion speed. The details of the deceleration are not known. Recent satellite measurements of sky background Lα-emission indicate an *interstellar wind* of neutral hydrogen flowing from the direction of Sagittarius and reaching deep into the solar system to within a few AU from the Sun. In a first approximation, we can consider the solar wind as if it were flowing through a uniform background of neutral gas with density 0.1 hydrogen atoms $\times \text{cm}^{-3}$. Charge exchange with this gas decelerates the solar wind and heats both, thus adding to the complexity of the solar wind termination.

It is likely that the heliosphere becomes quite elongated in the direction of flow of the neutral interstellar wind.

Axford, W. I.: 1973, *Space Sci. Rev.* **14**, 582.
Montgomery, M. D.: 1973, *Space Sci. Rev.* **14**, 559.

13.9. Sector Structure

Solar magnetic field lines are convected away from the Sun by the expanding solar wind. This *interplanetary magnetic field* is organized and ordered on a large scale. Observations made near the equatorial plane of the Sun show that the magnetic field is organized into a few (typically four) sectors or regions where the magnetic field is predominantly directed either away from the Sun or towards the Sun along the basic Archimedes spiral induced by solar rotation. The *sector boundary* separating fields of opposite polarity is normally very thin, being convected past the observer in the order of minutes as compared to the width of a typical sector of about a week. Strong indirect evidence leads to a concept of a *heliomagnetic neutral sheet* separating plasmas that flow from the two hemispheres of the Sun – which in general have oppositely directed magnetic fields. The two plasma regimes (northern and southern) are separated at the Sun by a broad equatorial belt of closed magnetic field lines from which little or no solar wind flows. Each time an observer traverses the neutral sheet he will note the passage of a sector boundary.

Global-scale azimuthal organization of the photospheric magnetic field causes a warping in interplanetary space of the basic equatorial heliomagnetic neutral sheet deflecting it alternately northwards and southwards from the equator (*warped current sheet*). Thus an observer near the solar equatorial plane detects plasma alternately from either side of the neutral sheet during the course of solar rotation. As a consequence a sector structure with alternating magnetic polarities is observed. Shortly after sunspot maximum, when the polar fields of the Sun weaken, the heliomagnetic neutral sheet may be very warped and irregular, but at most other times during the solar cycle the neutral sheet is well defined and the warping imposed by the photospheric sector structure is small. – When the Earth, in its orbit, approaches its maximum excursion from the solar equatorial plane (slightly more than 7°), the magnetic polarity corresponding to that of the nearer polar field dominates more and more, a phenomenon known as the *dominant polarity* or *Rosenberg-Coleman effect.* – Near sunspot minimum when the polar fields are strong, the warping can at times become negligible and a unipolar interplanetary magnetic field may be observed for several months at the Earth. The three-dimensional structure of the heliomagnetic neutral sheet is shown in Figure 13.1.

Rosenberg, R. L. and Coleman Jr., P. J.: 1969, *J. Geophys. Res.* **74**, 5611.
Schulz, M.: 1973, *Astrophys. Space Sci.* **24**, 371.
Svalgaard, L. and Wilcox, J. M.: 1976, *Nature* **262**, 766.
Wilcox, J. M.: 1968, *Space Sci. Rev.* **8**, 258.
Wilcox, J. M. and Ness, N. F.: 1965, *J. Geophys. Res.* **70**, 5793.

13.10. Spatial Gradients

Solar wind properties vary with distance from the Sun and with longitude. There is evidence for latitudinal variations as well. Some of these spatial variations are just consequences of the expansion into a larger and larger volume, others are the results of dynamical processes within the solar wind as it expands, while yet others reflect variations in the source conditions.

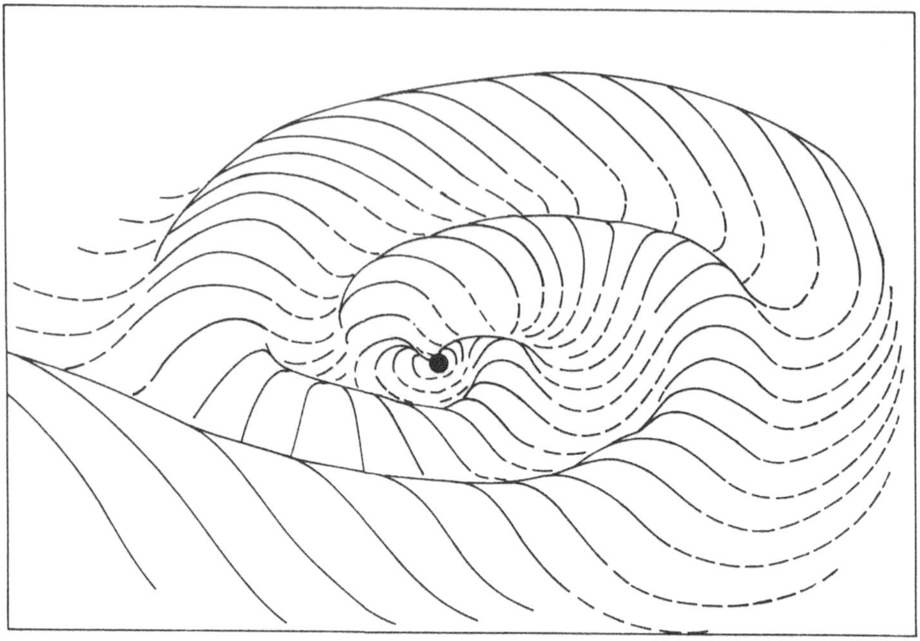

Fig. 13.1. Schematic diagram of the warped current sheet in the inner solar system (inside 6 AU). This current sheet divides the interplanetary magnetic field in the heliosphere into two regions with oppositely directed field lines. In one region the field polarity is toward the Sun, in the other region it is away from the Sun. The situation is shown for a four-sector structure. Full and dashed lines indicate the current sheet lying above and below the equatorial plane, respectively. The extent in latitude of the current sheet was assumed to be ±15°. The Sun at the centre is not shown to scale.

Out to at least ten AU the radial component of the interplanetary magnetic field as well as the solar wind density decrease simply inversely with the square of heliocentric distance consistent with conservation of the flux of mass and magnetic field lines. The average azimuthal component of the field is observed to decrease with distance R more rapidly than the $1/R$ dependence predicted for a uniform coronal expansion. It has been suggested that this strange effect is caused by a correlation between small-scale fluctuations in the flow speed and in the magnetic field. More observations are needed to confirm the existence of the effect. The average solar wind speed seems to be constant over the range covered by in-situ spacecraft measurements (0.3 to 10 AU) while the relative fluctuations in the flow speed decrease with larger distances.

Observations of scintillations of the radio flux from distant radio sources as the line-of-sight passes through the inner solar system allow measurement of solar wind speed at high heliographic latitudes. A small latitudinal gradient of the flow speed of +2 km s^{-1} per degree of latitude is consistent with the data.

Bame, S. J., Asbridge, J. R., Feldman, W. C., Felthauser, H. E., and Gosling, J. T.: 1977, *J. Geophys. Res.* **82**, 173.
Coles, W. A. and Rickett, B. J.: 1976, *J. Geophys. Res.* **81**, 4797.

Dobrowolny, M. and Moreno, G.: 1976, *Space Sci. Rev.* **18**, 685.
Neugebauer, M.: 1975, *Space Sci. Rev.* **17**, 221.

13.11. Long Term Variations

Solar wind speed and density measured near the ecliptic plane seem to be lower near sunspot maximum and higher near minimum, especially in the years just before the minimum. The total variation in both quantities is about 30% amounting to a difference in mass flux of a factor of two. It is possible that this sunspot cycle variation of the flux arises from variation in latitude of the source regions of the wind, since we tend to have a lesser flux of solar wind in the equatorial regions of the Sun due to the constraining forces of the magnetic fields of active regions.

The strength of the interplanetary magnetic field has been surprisingly constant over the last sunspot cycle with little or no change in yearly mean values.

Indirect evidence from geomagnetic activity indicates that the solar wind must have changed its properties considerably since about 1900 but it is not known what property (mass flux or magnetic flux) has changed. Because direct measurements extend just a little more than one sunspot cycle back in time, any long term variations are difficult to detect and none are firmly established.

Gosling, J. T., Asbridge, J. R., Bame, S. J., and Feldman, W. C.: 1976, *J. Geophys. Res.* **81**, 5061.
King, J. H.: 1976, *J. Geophys. Res.* **81**, 653.
Neugebauer, M.: 1975, *Space Sci. Rev.* **17**, 221.
Svalgaard, L.: 1977, in 'Coronal Holes', NASA SP.

13.12. Zodiacal Light

A faint glow extends away from the Sun nearly in the ecliptic plane. It is visible to the naked eye in the western sky shortly after sunset or in the eastern sky shortly before sunrise. Its spectrum indicates it to be sunlight scattered by interplanctary dust particles averaging about 1 micron in diameter, presumably representing the small-size part of the meteoritic size spectrum. The zodiacal light is also strongly polarized and was once thought to be due to Thompson scattering by free electrons. The scattering and polarization of light by particles with sizes near the wavelength of the light can be understood on the basis of Mie's scattering theory. This theory also predicts a 'backscatter peak' in the direction opposite the light source. A very faint glow – the *gegenschein* – is in fact observed in the antisolar direction forming a large diffuse patch which monotonically fades within $\pm 40^\circ$. Its optical spectrum confirms that the gegenschein is sunlight reflected from dust.

Leinert, C.: 1975, *Space Sci. Rev.* **18**, 281.
Roosen, R. G.: 1971, *Rev. Geophys. Space Phys.* **9**, 275.

14. SOLAR-TERRESTRIAL PHYSICS

V. L. PATEL

14.1. Geomagnetic Field

Magnetic observatories monitor continuously the geomagnetic field at various latitudes and longitudes of the Earth. With the advent of rockets and satellites, it has been studied also in space.

The geomagnetic field, to a first approximation, is equivalent to a dipole of magnetic moment M at the centre of the Earth (*central dipole*). This *equivalent dipole* represents the field more accurately if the dipole axis is displaced 451 km from the Earth's centre towards the Pacific Ocean (*eccentric dipole*). The dipole axis of the Earth is tilted approximately 11° to the Earth's rotation axis and is assumed to intersect the Earth's surface at 81° N, 84.7° W in Greenland, the *geomagnetic north pole*, and at 75° S, 120.4° E in Antarctica, the *geomagnetic south pole.* The position of the eccentric dipole needs to be adjusted from year to year in order to remain a good approximation to the observed geomagnetic field. The off-centre distance has increased from 252 km in 1829 A.D. to 451 km in 1965. The magnetic moment of the dipole also varies with time and has decreased systematically in the last hundred years, being given approximately by $M = (15.77 - 0.003951t) \times 10^{25}$ G cm^3 where t is the time in years from 1900 A.D. The dipole, central or eccentric, is only a crude approximation to the geomagnetic field. Geophysicists continuously modify and update theoretical models with higher multipoles and calculate spherical harmonic coefficients for several epochs. The International Association of Geomagnetism and Aeronomy (IAGA) publishes the International Geophysical Reference Field (IGRF) for the scientific community. The multipole coefficients for the 1975 IGRF field, including secular variations, are given by Leaton (1976).

The geomagnetic field intensity at any point is denoted by **F** or **B** and is measured in gauss (G or Γ), tesla (1 T = 10^4 G) or gamma ($1\gamma = 10^{-5}$ G = 1 nT).

Akasofu, S. I. and Chapman, S.: 1972, *Solar-Terrestrial Physics*, Oxford University Press, Ch. 2.
Haymes, R. C.: 1971, *Introduction to Space Science*, John Wiley and Sons, New York, Ch. 9.
Leaton, B. R.: 1976, *J. Geophys. Res.* **81**, 5163.

14.2. Coordinate Systems

The geomagnetic field intensity **B** is represented in a *local geomagnetic coordinate system* by B, the magnitude of the vector, and two angles, I and D. The angle I is the angle between **B** and its horizontal component H; it is positive downward and is called the *inclination* or the *dip angle.* The *declination* D is the azimuth from the northward horizontal direction measured either eastward or westward. At most magnetic observatories D is measured positive eastward. The field **B** can also be given by three perpendicular components: X northward, Y eastward in the horizontal plane and Z or V

Bruzek and Durrant (eds.), Illustrated Glossary for Solar and Solar-Terrestrial Physics. 159–193.

vertically downward (positive if downward). The seven quantities B, H, D, I, X, Y, Z or V are called *magnetic elements*. Figure 14.1 shows the inter-relationship of these quantities. Several magnetic observatories record these elements continuously. The records, called *magnetograms*, are stored in World Data Centers for future use.

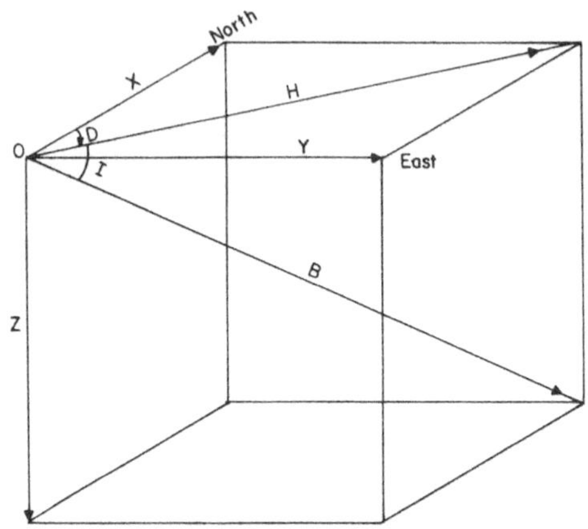

Fig. 14.1. Local coordinate system used for the geomagnetic field. The angle I is the inclination or dip and the angle D is the declination. The X axis is towards the north, Y towards the east and Z towards the centre of the Earth.

When $I = 0$, the field is horizontal; the points at which $I = 0$ define a curve on the surface of the Earth called the *dip equator.* The inclination or dip angle I is positive north of this equator and negative south of it. When the horizontal component H is zero i.e. $I = 90°$, the field is vertical; this occurs at two points called the *north dip pole* and *south dip pole.* For the epoch 1965 the dip poles were at 75.6° N, 101° W and 66.3° S, 141° E. The positions of the dip poles change reflecting internal changes of the Earth's field. The main geomagnetic field for a given epoch year is represented by *isomagnetic maps* showing *isomagnetic* (constant value) *lines* for each magnetic element. In the literature, the isomagnetic lines for D are called *isogonic lines* or *isogones*, for I, *isoclinic lines* or *isoclines* and for B, *isodynamic lines*. For other elements such as H, Z, X and Y, they are simply called isomagnetic lines or *equilines*.

Another coordinate system that is widely used in geomagnetic research is that of *geomagnetic coordinates* or *dipole geomagnetic coordinates.* The *geomagnetic latitude* and *geomagnetic longitude* of a point on the Earth are calculated with respect to a central dipole.

The geographic latitude and longitude of the north pole of the central dipole are denoted by θ_0 and Φ_0 respectively. For example, the epoch 1965 had $\theta_0 = 78.6°$ N (colatitude 11.4°) and $\Phi_0 = -69.8°$ or 290.2° E.

A point with geographic latitude θ and longitude Φ has geomagnetic latitude λ_m and geomagnetic longitude Λ given by

$$\sin \lambda_m = \cos \theta_0 \; \cos \theta \; \cos(\Phi - \Phi_0) + \sin \theta_0 \; \sin \theta$$

$$\cos \Lambda = \frac{\sin \theta_0 \; \cos \theta \; \cos(\Phi - \Phi_0) - \cos \theta_0 \; \sin \theta}{\cos \lambda_m}$$

It is also possible to calculate *dip latitude* λ_D with respect to the dip-pole and the dip-equator. In the central dipole model, the dip angle I and dip latitude λ_D are related by $\tan \lambda_D = \frac{1}{2} \tan I$.

There are modifications and corrections to the dipole coordinates since the Earth's magnetic field is not a pure dipole (see Cole, 1963; and Gustafsson, 1970).

In research literature, *dipole time* or *geomagnetic time* is used in studies of auroras and high-latitude phenomena. The geomagnetic time for a point on the Earth having geographic coordinates θ, Φ and geomagnetic coordinates λ_m, Λ, is calculated from the dipole noon,

$$t = \frac{\Lambda_H - \Lambda}{15^\circ},$$

where Λ_H is the dipole longitude of the Sun and t is in hours.

Akasofu, S. I. and Chapman, S.: 1972, *Solar-Terrestrial Physics*, Oxford University Press, Ch. 2.
Cole, K. D.: 1963, *Australian J. Phys.* **16**, 423.
Gustafsson, G.: 1970, *Ark. Geofis.* **5**, 595.

14.3. Geomagnetic Indices

The geomagnetic field shows fluctuations on various time scales. To describe them, several geomagnetic indices have been devised for statistical studies in solar-terrestrial physics.

The most widely-used index for geomagnetic disturbances is the *planetary index Kp* which is derived from the K index of a number of selected magnetic observatories. The K *index* at a given observatory is obtained by inspecting the H, D, Z elements for the eight daily intervals of three hours 00–03, 03–06 UT etc. The largest range of H, D and Z for each interval is used to obtain the K index from a standard table. The table for each station is adjusted so that the largest and the smallest variations occurring at this station are represented by $K = 9$ and $K = 0$ respectively. The K index is on a semi-logarithmic scale. The 3-hr Kp index is obtained by taking a mean of the K indices measured at twelve observatories. The planetary 3-hr Kp index is usually expressed in one-third units by adding the signs $-$, $_0$, $+$ to the numbers 0 to 9; that gives a 28-step scale from 0_0 to 9_0. The Kp values are published by the Göttingen Institute for Geophysics in the form of Bartels's *musical diagram* (see *Solar-Geophysical Data*, Prompt Reports, NOAA Boulder, Col.). The daily ΣKp is the sum over the eight 3-hr indices of a universal day. Although the Kp index is widely used in solar-terrestrial physics, its physical meaning is unclear because of the limited global data used in its derivation.

Another index recently used in solar-terrestrial physics is the *ap index* which is a linear index derived from each 3-hr planetary index Kp. The conversion is such that at 50°

dipole latitude *ap* approximately represents the maximum disturbance range of the largest of the three elements *H, D* and *Z* in units of 2.0 nT. The conversion table is given in Akasofu and Chapman (1972). The *ap* index ranges from 0–400. The sum of eight 3-hr *ap* values gives the daily *Ap index* for a universal day.

The lesser-used *index Cp*, the *daily planetary character index*, is obtained from the daily *Ap* index. It has values from 0 to 2.0. The *index C9*, which has a range of 0 to 9, is derived from the daily *Cp* index and usually is tabulated with the sunspot number *R*.

The indices *C* and *Ci* are not widely used. The *index C* is a daily character index assessed at an observatory by simple inspection. The *C* index has the values 0, 1 or 2. The *international C index Ci* is obtained by taking averages (to one decimal place) of several observatories.

Recently, Mayaud (1972) proposed new indices in which daily or yearly variation effects are removed. He defined the index *aa* (*a* for antipodal) as the average of 3-hr *K* indices converted to the amplitude of the field at two antipodal observatories. The Greenwich and Melbourne observatories are roughly antipodal and have data since 1867. The *aa* index is similar to the *ap* index. The *daily index Aa* is obtained from the eight *aa* indices. The *Aa* indices are more convenient than *Ci* because the variations appear larger in time-series studies. The 3-hr *am* and the daily *Am indices* also are derived from four or five groups of northern and southern observatories. The difference between *aa* and *am* indices rapidly disappears when averages of several 3-hr values are considered. Mayaud has also devised the *Km index* as a recalibrated and improved *Kp* index by using better distributed observatories. However, the index *aa* is a better index for long-term studies.

There are three-hourly *Kn, Ks, an, as* indices for northern and southern hemispheres derived from *K* indices of eleven northern and seven southern subauroral zone observatories.

Akasofu, S. I. and Chapman, S.: 1972, *Solar-Terrestrial Physics*, Oxford University Press, Ch. 7.
Haymes, R. C.: 1971, *Introduction to Space Science*, John Wiley, New York, Ch. 9.
Mayaud, P. N.: 1972, *J. Geophys. Res.* **72**, 6870.
Rostocker, G.: 1972, *Rev. Geophys. Space Phys.* **10**, 935.
Svalgaard, L.: 1976, *J. Geophys. Res.* **81**, 5182.

14.4. *Dst* Index

Geomagnetic storms are fluctuations in the main geomagnetic field that show decreases of several hundred nT in the *H* component for several days. The *storm vairation D* is defined as $D = Dst + DS$, where *DS* is due to auroral electrojet activity and *Dst* is assumed to be primarily due to the ring current in the magnetosphere. At present the *storm-time variation index Dst* is calculated using the *H* component of four observatories: Honolulu, San Juan, Kakioka and Hermanus which are uniformly distributed in longitude and are away from the effects of the auroral and equatorial electrojets. The method of derivation of the *Dst* index is given by Sugiura. The use of the *Dst* index is not limited to geomagnetic storms. The continuous hourly values are widely used to characterize quiet-time geomagnetic variations for solar-terrestrial studies.

Patel, V. L. and Desai, U. D.: 1973, *Astrophys. Space Sci.* **20**, 431.
Rostocker, G.: 1972, *Rev. Geophys. Space Phys.* **10**, 935.
Sugiura, M.: 1964, *Annals IGY*, Vol. XXXV, Pergamon Press, p. 9.

14.5. Auroral Electrojet Index *AE*

The geomagnetic indices *Kp* or *Ap* and *Dst* give information on planetary disturbances and geomagnetic storm variations. The currents in the ionosphere of the auroral zone have a marked effect on the geomagnetic field. The magnetic field fluctuations due to these auroral-zone currents (auroral electrojet) are represented by the *AE* index. In order to calculate the *AE* index, the *H* component from observatories located in auroral or sub-auroral latitudes and uniformly spaced in longitude is needed. In the past, the observatories used were located between 60° and 64° N plus three in the southern hemisphere at −63.8°, −65.8° and −70.6° to fill the gaps in the longitude range. The perturbation ΔH is obtained every 2.5 min by using a quiet-time base-line for *H* at each station. Then ΔH is superposed for all stations.

The maximum positive (upper) amplitude of *H* is called the *AU index* and the maximum negative (lower) amplitude is called the *AL index*. Physically, the *AU* index is the maximum magnetic perturbation due to the eastward electrojet current in the afternoon sector and *AL* is the maximum magnetic perturbation effect due to the westward electrojet current in the morning sector. Sometimes, *AU* and *AL* indices are used independently because evening and morning electrojet currents may flow independently. In solar-terrestrial studies, the index $AE = AU - AL$ is used; it represents the difference between the upper and lower limits of the magnetic fluctuations in ΔH at any time. The *AE* index is given in gamma or nanotesla at either 2.5 min intervals or more usually at hourly intervals and is available from the *World Data Center*, NOAA, Boulder, Colorado.

Davis, J. N. and Sugiura, M.: 1966, *J. Geophys. Res.* **71**, 785.
Rostocker, G.: 1972, *Rev. Geophys. Space Phys.* **10**, 951.

14.6. Daily Variations of Geomagnetic Field

The total magnetic field intensity of the Earth is divided into several parts, $B = M + m + D + N + S + L$ where *M* is the main field arising from internal sources, *m* is the *secular variation, D* is the regular or irregular *disturbance field* mainly due to external sources, *N* is the non-cyclic variation in quiet periods, also from external sources. The *S field* is periodic and arises from the influence of the Sun. The *L field* is also a periodic field perturbation, due to the influence of the Moon.

Days during which the magnetic field records (magnetograms) of three elements *H, D, Z* or *X, Y, Z* show regular smooth variations are called magnetically *quiet days.* If large irregular fluctuations occur, they are called *disturbed days.* The solar magnetic variation *S* on quiet days is called *Sq.* The daily variation on disturbed days after exclusion of *Sq* is called the *disturbance daily variation* S_D. It is part of the world-wide disturbance field *D.* In order to study 'true' *Sq* variations, all other variations must be removed using statistical methods. The *Sq* variation is explained in terms of solar tidal motions (S_1) and thermo-tidal motions (S_2) of the atmosphere causing ionospheric motions. The ionosphere is a conducting plasma layer and its motion induces a Lorentz force on the ions and electrons which, in turn, sets up electric currents in the ionosphere (fixed with respect to the Sun). Similarly, the lunar tide causes a Lorentz dynamo force and an associated current system fixed with respect to the Moon. The station moving under these current systems will observe the lunar-time dependent magnetic variations of these currents. They are also called simply *L-variations* and have semi-diurnal characteristics with maxima at 6 and 18 lunar hours. From surface observations of the geomagnetic field the *Sq* and *L* variations can be isolated and idealized equivalent current systems in the ionosphere can be deduced. A typical *Sq* variation is about 20 nT at mid-latitudes and is clearly visible on magnetic records, while *L* is very small (few nT) and needs statistical analysis to extract it.

The solar daily variation is enhanced near the dip-equator and may be 100–200 nT. The *Sq* current system is enhanced by the electrojet current flowing in a narrow region ($\pm 5^\circ$) over the equator. The *Sq* current system is a vortex of currents located in each hemisphere centred around 30° geomagnetic latitude and near noon meridian. The currents flow around this centre, called the *Sq focus*, in a clockwise direction in the southern hemisphere and anti-clockwise in the northern hemisphere (the north and south *Sq* foci are not on the same meridian). The strong *Sq* currents are in the sunlit side and between the equator and midlatitudes. The equivalent currents may be ≈ 120 000 ampères. Rocket experiments carrying magnetometers have been undertaken at middle latitudes to detect these currents. The current system changes from day to day, intensifies in summer and is affected by solar flares (solar flare effect Sfe), by a solar eclipse, and possibly by the interplanetary field and magnetospheric dynamics.

There is another current system invoked by Nagata and Kokubun to explain the daily variation in the polar cap region on quiet days. The two-vortex system, called S_q^p, is obtained after subtracting the *Sq* variation arising from dynamo action; it is confined to high-latitude regions within 60° geom. latitude. Recent results indicate that on quiet days the S_q^p *currents* are mainly confined to the high-latitude 06–12 and 12–18 hour quadrants.

Akasofu, S. I. and Chapman, S.: 1974, *Solar-Terrestrial Physics*, Oxford University Press, Ch. 4.
Kane, R. P.: 1976, *Space Sci. Rev.* **18**, 413.
Matsushita, S.: 1967 in S. Matsushita and W. H. Cambell (eds.), *Physics of Geomagnetic Phenomena*, Vol. I, Ch. 3, Academic Press, New York.
Nagata, T. and Kokubun, S.: 1962, *Rep. Ionosph. Space Res. Japan* **16**, 256.
Price, A. T.: 1969, *Space Sci. Rev.* **9**, 151.

14.7. 27-Day Variations of Geomagnetic Field

Geomagnetic indices *Ci, Kp* and *Ap* show 27-day cycles in geomagnetic activity of solar origin. The active regions on the Sun emit solar plasma in the form of high-speed streams which cause geomagnetic disturbances. Since active solar regions sometimes persist for several solar rotations they give rise to a 27-day periodicity in geomagnetic activity. Several solar-terrestrial phenomena (e.g. auroral *Sq*, earth-current, cosmic ray intensity, aurora, etc.) reflect the 27-day periodicity of geomagnetic activity, with increased amplitude during periods of high geomagnetic activity. There are several subperiods of 12 to 14 and 6 to 9 days observed in geomagnetic activity. It is possible that these smaller periodicities are generated by the interplanetary magnetic field sectors interacting with the geomagnetic field. In solar-terrestrial studies they are apparent in the cosmic-ray intensity and in interplanetary field sectors.

The amplitude of the 27-day wave appears to be modulated by the semi-annual variation of geomagnetic activity. There is no significant modulation of the 27-day wave amplitudes by the annual or the 11-yr period.

Chapman, S. and Bartels, J.: 1940, *Geomagnetism*, Vol. I., Oxford University Press, Chap. 12.
Patel, V. L. and Chasson, R. L.: 1968, *Can. J. Phys.* **46**, S966.
Shapiro, R., 1969, *J. Geophys. Res.* **74**, 2356.
Wilcox, J. M. and Ness, N. F.: 1965, *J. Geophys. Res.* **70**, 5793.

14.8. Semi-Annual Variation of Geomagnetic Field

Monthly means of the geomagnetic indices *Ci* have been used to study annual variations. There is some controversy about the existence of a significant annual (12 month wave) period, but there are two prominent peaks near March 22 and September 20. This semi-annual variation of about 15 nT amplitude (the amplitude varies directly with solar activity) with maxima near the equinoxes and minima near the solstices is now well established. However, the physical mechanisms are still not definitely known.

There are two older theories. One uses the axial hypothesis based on the Earth's heliographic latitude: the Earth is advantageously located to interact with solar plasma streams around September 7 when its heliographic latitude is a maximum north 7.2° and around March 6 when it is a maximum south 7.2°. The other theory uses the equinoctial hypothesis based on the tilt of the equivalent geomagnetic dipole axis. It is assumed that the maximum geomagnetic activity results when the dipole axis is perpendicular to the solar wind plasma flow which occurs near the equinoxes. Recently, Boller and Stolov have examined the role of Kelvin-Helmholtz instabilities at the magnetopause in order to explain the semi-annual variation. Using magnetic field and plasma data in space, they argue that the Kelvin-Helmholtz instability has a maximum near the equinoxes and a minimum near the solstices. The magnetic activity is modulated accordingly. The K-H instability also predicts universal time variations which they verify in geomagnetic data using *K* and *Kp* indices. The model depends on the tilt of the dipole axis and the maximum and minimum probabilities of the K-H instability thus depend on the season.

The geomagnetic field has also long periodic variations. A small annual variation is reported and, though not well established, can be explained by ionospheric winds. Moreover, there are well-known 11- and 22-yr periods related to the sunspot cycle. Other periods range from 2 yr to 80 yr (see Currie, 1973; and Kane, 1976).

Boller, B. R. and Stolov, H. L.: 1970, *J. Geophys. Res.* **75**, 6073.
Currie, R. G.: 1973, *Astrophys. Space Sci.* **21**, 425.
Currie, R. G.: 1976, *J. Geophys. Res.* **81**, 2935.
Kane, R. P.: 1976, *Space Sci. Rev.* **18**, 413.
Mishin, V. M., Kalinovskaya, G., and Kishina, N. A.: 1961, *Geom. Aeron.* **1**, 347.
Shapiro, R.: 1969, *J. Geophys. Res.* **74**, 2356.

14.9. Micropulsation

The geomagnetic field exhibits fluctuations with periods from a fraction of a second to

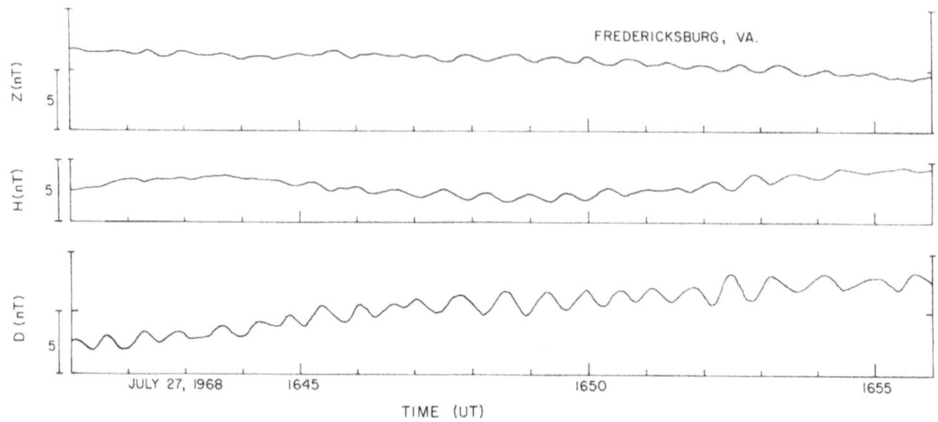

Fig. 14.2. A typical event of micropulsation (Pc3) at Fredericksburg, Virginia magnetic observatory.

tens of minutes. These variations have small amplitudes ranging from a tenth of a nT to several hundred nT in high latitudes. They are of a transitory nature, lasting for a few minutes to an hour or more. These fluctuations are commonly called micropulsations or simply pulsations or geomagnetic micropulsations. An example is shown in Figure 14.2.

In the older literature, terms like continuous pulsations (Pc), pulsation train (PT), giant pulsations (Pg) have been used. Recently, geomagnetic pulsations have been classified into two main categories: Pc and Pi. *Pc1 pulsations* with periods 0.2 to 5 s may occur in the form of bursts, developing as a series of bursts that may last from minutes to hours. The Pc1 are called *pearls*. The Pc2 to Pc5 pulsations have a regular, continuous, almost sinusoidal pattern while *Pi pulsations* have an irregular pattern in rapid-run magnetograms. The classification presently in use is as follows:

Pc1	period	0.2–5 s		Pi1	period	1–40 s
Pc2		5–10 s		Pi2		40–150 s
Pc3		10–45 s				
Pc4		45–150 s				
Pc5		150–600 s				

The micropulsations occur at magnetically quiet as well as at disturbed times. However, Pi's are more likely to occur under disturbed conditions such as substorms. The occurrence of Pi2 has been associated with the onset of substorms.

The short period pulsations Pc1 are studied by using frequency-time diagrams called *sonagrams*. These are sensitized papers which show blackness proportional to the signal strength at a given frequency and time. Theories have been proposed that Pc1 and Pi1 originate from cyclotron instabilities, wave particle interactions and bouncing motion of particles along the magnetic field lines in the magnetosphere.

The short-period *Pi1 pulsations* are a very complex phenomenon and are divided into subclasses:

(1) SIP is a train of short-period micropulsations which lasts only for several minutes (*short irregular pulsation*). They usually occur near the onset of geomagnetic bays and are

associated with 20–60 s long-period pulsations. They tend to repeat a few times at intervals of 10–20 min.

(2) IPDP: *Irregular pulsations of diminishing period* have narrow frequency bands whose mean frequency increases from 0.5 to 1.5 Hz in 30 min. They usually occur from 1700 to 0100 local magnetic time and are often associated with the appearance of negative magnetic bays. They also occur with visual auroras. The frequency of IPDP increases at the time of auroral sporadic *E* appearance.

(3) IPIP: *Irregular pulsations of increasing period* usually occur in daytime and are not observed frequently.

(4) AIP: *Auroral irregular pulsations* occur near the auroral zones. They are continuous broad-band signals of long duration, usually preceded by SIP events.

There are many studies of the correlations of micropulsations with solar-terrestrial phenomena such as *Kp* index, solar cycle, seasons, interplanetary plasma etc. with, however, contradictory results. Generation mechanisms and propagation characteristics of micropulsations are still the subject of research; one interpretation is that the pulsations are hydromagnetic waves originating in the magnetosphere. Strong evidence for the magnetospheric origin was provided by Explorer 12 satellite observations. The micropulsation has latitude conjugacy in the southern and northern hemisphere of the Earth indicating field oscillation in the magnetosphere.

Jacobs, J. A.: 1970, *Geomagnetic Micropulsations*, Springer Verlag, New York, Ch. 2.
Lanzerotti, L. J. and Fukunishi, H.: 1974, *Rev. Geophys. Space Phys.* **12**, 2193.
McPherron, R. L., Russell, C. T., and Coleman, P. J., Jr.: 1972, *Space Sci. Rev.* **13**, 411.
Patel, V. L.: 1965, *Planet. Space Sci.* **13**, 485.
Patel, V. L. and Cahill, L. J., Jr.: 1964, *Phys. Rev. Letters* **12**, 213.

14.10. Geomagnetic Storms

Large decreases (order of 100 to several hundred nT) in the *H* component of the geomagnetic field sometimes occur simultaneously at middle and low latitudes on the Earth and are called geomagnetic or magnetic storms. A typical magnetic storm perturbation at Honolulu (geom. lat. 21° N) is shown in Figure 14.3. The common features of magnetic storms are: sudden commencement (SC) or storm sudden commencement (SSC), initial phase (IP), main phase (MP) and recovery phase (RP). The storm begins with a sharp increase in the *H* component at all latitudes. The increase is referred to as SSC or SC. For several hours the field remains above pre-SC values until a sudden decrease begins. The time interval between SC and the onset of the decrease is called the *Initial Phase.* The decrease is seen very clearly at middle and low latitudes. However, there are large superposed perturbations at equatorial stations due to the electrojet. The decrease may last several hours or sometimes days and is called the *Main Phase.* Then slow recovery starts and the *H* component returns to the pre-storm level or in many cases to slightly below this level. This phase of the storm, called the ***Recovery Phase***, may last from hours to several days. The physical processes involved in these phases of storms are discussed under separate headings. Note that storms are well defined at middle latitudes but in the auroral zone the storm-related variations are quite different due to the auroral electrojet. The storm variations in higher latitudes beyond the auroral

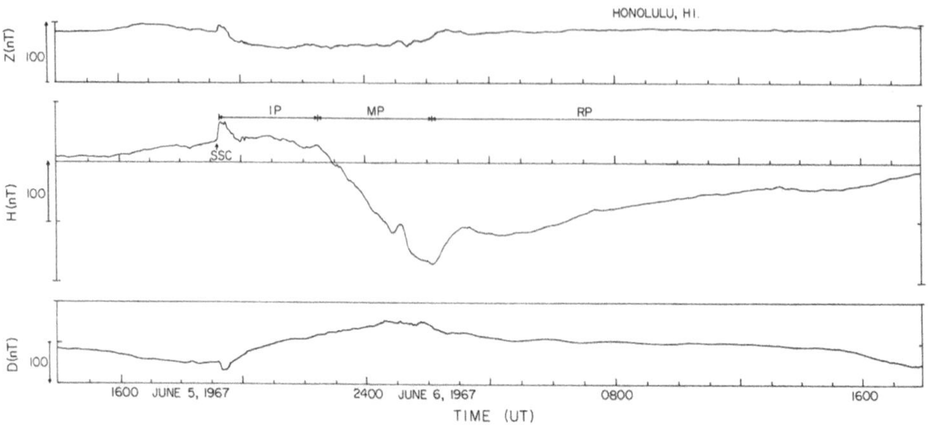

Fig. 14.3. A typical geomagnetic or magnetic storm variation in the geomagnetic field observed at mid-latitude station, Honolulu, Hawaii on June 5–6, 1967.

zone (in the polar cap) are again different because the interplanetary field may reconnect with the geomagnetic field and directly influence the magnetic variations.

The storms are usually caused by the arrival of high-speed solar plasma (generated by a solar flare) and the associated shockwave near the Earth. Some of the geomagnetic storms have a 27-day recurrence due to the return of active solar regions after one solar rotation. It should be noted that storm-time decreases are observed without SSC and also in many irregular forms. However, in recent years, the term magnetic storm has been used with the understanding that it has all the characteristic phases described: SSC, IP, MP and RP. Geomagnetic storms are also associated with ionospheric disturbances which are known to cause disruption in radio and television communications.

Akasofu, S. I. and Chapman, S.: 1974, *Solar-Terrestrial Physics*, Oxford University Press, Ch. 8.
Sugiura, M. and Heppner, J. P.: 1968 in W. N. Hess and G. D. Mead (eds.), *Introduction to Space Science*, Gordon-Breach Science Publ. Ch. 1.

14.11. Storm Sudden Commencement (SSC) or Sudden Commencement (SC)

The SSC is a sudden increase in the *H* component of the geomagnetic field which marks the beginning of a geomagnetic storm. The SSC occurs almost simultaneously all over the Earth. It has a tendency to occur one or more minutes earlier in high latitudes than in low latitudes. The increase starts suddenly but has a rise time of 150–300 s (Figure 14.3). The rise time of SSC depends on the local time of observation being 150 s near noon and 300 s around midnight. The magnitude of the SSC increase may range from 20–30 nT in low middle latitudes to hundreds of nT near the equator and in high latitudes. It has a large daily variation (largest amplitude near local noon) at stations near the magnetic dip-equator. For example, the ratio of the SSC variations at Huancayo (dip lat. 0.6°) and Fredericksburg (dip lat. 49.6°) is about 6 at noon and unity between 1600 and 0400 local time. The polarization of the SSC disturbance vector constructed from *H* and *D* components shows clockwise or counter-clockwise rotations depending on latitude and local time of observation. There is an indication that the local time-dependence of the magnitude and polarization characteristics are similar in the magnetosphere.

Occasionally, one observes a negative impulse before the main positive increase of the SSC. Such negative changes are called *SSC** or *primary reverse impulses* (PRI). The SSC* usually occurs in high latitudes or near the equator. Sometimes there are micropulsations associated with the SSC. The PRI have been explained on the basis of propagation and interaction of isotropic and transverse hydromagnetic waves in three dimensions. It is now established from interplanetary field and plasma data that the SSC is produced by the interaction of the hydromagnetic shockwave, associated with solar flares, with the magnetosphere. However, the actual mode of interaction is not yet established in every detail. In solar-terrestrial studies, it has been noted that there are SSC-associated disturbances in the ionosphere, in the occurrence of micropulsations and particle precipitation at high latitudes.

Akasofu, S. I. and Chapman, S.: 1974, *Solar-Terrestrial Physics*, Oxford University Press, Ch. 8.
Barcus, J. R.: 1972, *Space Sci. Rev.* **13**, 295.
Burlaga, L. F. and Ogilvie, K. W.: 1969, *J. Geophys. Res.* **74**, 2815.
Ondoh, T.: 1963, *J. Geomagn. Geoelect.* **14**, 198.
Patel, V. L. and Cahill, L. J., Jr.: 1974, *Planet. Space Sci.* **22**, 1117.
Tamao, T.: 1964, *Rep. Ionosph. Space Res. Japan* **18**, 16.
Wilson, C. R. and Sugiura, M.: 1961, *J. Geophys. Res.* **66**, 4097.

14.12. Initial Phase (Geomagnetic Storm)

In storm-time geomagnetic disturbances, the interval between the SSC and the onset of the main-phase decrease in the *H* component is defined as the Initial Phase (IP). The length of the IP is an important parameter from the theoretical point of view because the SSC signifies the arrival of a shock wave from interplanetary space and the Earth is submerged in post-shock plasma and field during the IP. Statistical studies indicate that the IP may last from 30 min to several hours.

Typically, the geomagnetic field at low latitudes is higher than the pre-SSC value and high-latitude stations show increased and highly irregular fluctuations during the IP. The magnetosphere is subject to large fluctuations of post-shock plasma and field on the order of a few seconds to hours time scale. The physical significance of the post-shock plasma surrounding the Earth is not yet fully understood.

Antisilevich, M. G.: 1967, *Geomagn. Aeron.* **7**, 265.
Patel, V. L. and Wiskerchen, M. J.: 1975, *J. Geomagn. Geoelect.* **27**, 363.

14.13. Ring Current

The *main phase* of the geomagnetic storm represents the decrease in the *H* component at middle and low latitudes. Typical decreases are 50 to 400 nT lasting a few hours to more than a day. The largest *Dst* index gives fairly accurately the magnitude of the main phase of a storm. The explanation of the main phase decrease is that reconnection of the interplanetary field with the geomagnetic field and the fluctuating size of the magnetosphere (which depends on the solar plasma momentum flux) allow the entry of new particles into the magnetosphere or the acceleration of the ambient plasma to keV energy range thus forming a *ring current* near 3 to 5 Earth-radii. This ring current, flowing westward in the vicinity of the Earth, generates a magnetic field opposite to the geomagnetic field thus decreasing it.

Recent experiments on satellites firmly establish that the ring current exists near 3 to 5 R_E during the main phase. Protons in the energy range 10–120 keV form a *storm-time ring current.* Even during quiet-time (without storms) trapped protons of high energy (300–872 keV) form a *quiet-time ring current* near 2.5 to 4 R_E. The ring current is the effective current generated by protons drifting westward and electrons drifting eastward in the dipole field of the Earth. The drift motion is mainly due to a ∇B force; the net effect is a westward current.

The *recovery phase* of geomagnetic storms starts when the ring current begins to decay, i.e. when the trapped particles begin to diffuse in the magnetosphere. The principal mechanism for the loss of particles is coulomb scattering and charge exchange of protons with neutral hydrogen, e.g. $H^+ + H \rightleftharpoons H + H^+$. Recently several plasma instabilities have been invoked to explain the decay as well as micropulsations observed during the existence of the ring current.

Akasofu, S. I.: 1963, *Space Sci. Rev.* **2**, 91.
Cahill, L. J., Jr.: 1966, *J. Geophys. Res.* **71**, 4505.
Dessler, A. J. and Parker, E. N.: 1959, *J. Geophys. Res.* **64**, 2239.
Frank, L. A.: 1967, *J. Geophys. Res.* **72**, 3753.
Sckope, N.: 1966, *J. Geophys. Res.* **71**, 3125.
Smith, P. H. and Hoffman, R. A.: 1973, *J. Geophys. Res.* **78**, 4731.

14.14. Sudden Impulses (Geomagnetic) SI

The *H* component of the geomagnetic field shows sudden variations of several nT magnitudes at low latitudes. These sudden perturbations may be increases or decreases appearing almost simultaneously all over the Earth. They are called sudden impulses SI^+ (increase) and SI^- (decrease). The difference between positive SI^+ and the storm sudden commencement SSC is that the SI is not followed by a geomagnetic storm. The negative SI^- is usually less abrupt than the SI^+. In the literature SI^+ and SSC are sometimes simply classified as SI. These sudden changes last for an hour or so and then the field intensity returns to the normal level. In higher latitudes similar disturbances appear simultaneously with those of low latitudes. A detailed study of SI's shows that the diurnal variations of amplitudes (similar to *Sq*) and simultaneity of occurrence are similar to the characteristics of SSC. Some of the SI^+s are caused by interplanetary shocks of solar origin. The negative SI^- are mostly associated with interplanetary hydromagnetic discontinuities.

The magnetosphere shows compression or rarefaction during a SI perturbation, the amplitude of the SI having similar diurnal variations in the magnetosphere and on the surface of the Earth. The SI variations have been found to propagate in the magnetospheric tail with a speed of 700 to 1000 km s^{-1}. Because SI's are not associated with intense storm variations, they offer the unique opportunity to study the propagation of non-linear perturbations in low density plasma (that can not be generated in the laboratory). The SI's may have a significant effect on the life-time of the trapped radiation in the magnetosphere, ionosphere and the particle precipitation in the auroral zones.

Nishida, A. and Cahill, L. J., Jr.: 1964, *J. Geophys. Res.* **69**, 2243.
Nishida, A. and Jacobs, J. A.: 1962, *J. Geophys. Res.* **67**, 525.
Patel, V. L.: 1968, *J. Geophys. Res.* **73**, 3407.
Patel, V. L.: 1972, *Planet. Spac Sci.* **20**, 1127.
Patel, V. L. and Coleman, P. J., Jr.: 1970, *J. Geophys. Res.* **75**, 7255.
Patel, V. L. and Cahill, L. J., Jr.: 1974, *Planet. Space Sci.* **22**, 1117.

14.15. Ionosphere

The atmosphere of the Earth is divided into several regions according to the thermal and chemical properties of the neutral gas and ionized components. Various names are assigned to the various parts of the atmosphere as shown approximately in Figure 14.4.

The ionosphere is that part of the upper atmosphere where ions and electrons of thermal energy are present in sufficient quantity to modify the electromagnetic wave propagation. The ionization of the neutral atmospheric constituents is caused by the solar

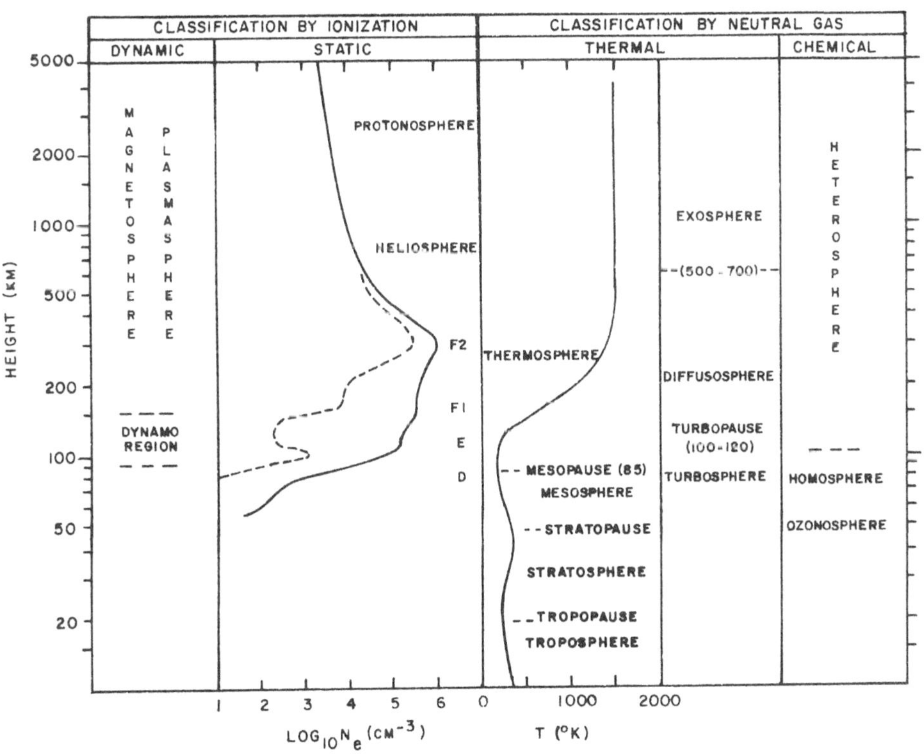

Fig. 14.4. Classification of the atmospheric regions of the Earth based on the ionization, the thermal and chemical properties of the atmospheric gas.

radiation impinging on the atmosphere; therefore, the electron density in the ionosphere varies with solar zenith angle, solar activity such as sunspots, solar cycle, the time of the day and seasons. The ionosphere for any planet is defined as that part of its atmosphere where free electrons and ions of thermal energy exist under the control of the gravity and magnetic field of the planet. The terrestrial ionosphere is divided into four layers D, E, F_1 and F_2 as shown in Figure 14.4. The actual heights of these layers vary with time, seasons and solar activity and therefore it is difficult to set definitive values. The figure shows the average behaviour of the atmosphere.

The electron densities in the E and F layers are routinely obtained by *ionosondes.* An ionosonde is a transmitter sending frequency-modulated pulses which are refracted from the ionosphere at the level where $f = f_p$. (The term 'reflected' is not appropriate although commonly used in the literature). The electron density N_e is derived from the plasma frequency f_p : N_e (cm^{-3}) = 12 400 f_p^2 (MHz).

The maximum plasma frequency of the E layer is called f_0E and similarly for the other layers f_0F_1 and f_0F_2. The corresponding maximum electron densities are denoted by N_mE, $N_mF_{1,2}$. The traces showing approximately virtual height vs frequency (i.e. electron density) are called *ionograms.* More than one hundred observatories prepare ionograms on a routine basis to monitor the characteristics of the ionospheric layers. The observations have been made also by satellite experiments from above the layers (*topside sounders*).

Bauer, S. J.: 1973, *Physics of Planetary Ionospheres*, Springer Verlag, New York, Ch. 1.
Chapman, S.: 1950, *J. Atmos. Terr. Phys.* **1**, 121.
Risbeth, H. and Garriot, O. K.: 1969, *Introduction to Ionospheric Physics*, Academic Press, Ch. 1.

14.16. *D* Region

The D region of the ionosphere is approximately between 50–90 km (Figure 14.4), however the exact heights are difficult to define. The region between 50–70 km is sometimes called the *C region* or *lower D region.* The D region electron densities are small and the radio wave absorption is quite significant. For these reasons, measurements of electron density are difficult to make and thus involve large uncertainties. Electron densities in daytime may be 10^2–10^4 cm^{-3} depending on height. Daytime electron densities show diurnal, seasonal and sunspot (solar activity) dependence. Although rocket measurements during night-time indicate small electron densities at D region level the existence of a night-time D layer is doubtful.

The ionizing agents of the D layer are still not clearly understood but solar Lα (1216 Å) radiation seems to be important in ionizing NO (nitric oxide). Rocket and satellite observations indicate that solar cosmic rays and primarily solar protons of 1–100 MeV and possibly solar electrons of energy >10 keV contribute to the ionization of the D layer.

The composition of the D layer is subject to research. NO^+ is believed to be the major positive charge carrier, and electrons, O_2^- and possibly other negative ions are negative charge carriers. The neutral component of the D region consists of mainly N_2, O_2, Ar, CO_2, He and a highly variable quantity of O_3 and H_2O.

Bauer, S. J.: 1973, *Physics of Planetary Ionospheres*, Springer Verlag, New York, Ch. 9.
Reagan, J. B. and Watt, T. M.: 1976, *J. Geophys. Res.* **81**, 4579.
Sechrist, C. F., Jr.: 1975, *Rev. Geophys. Space Phys.* **13**, 894.
Van Zandt, T. E.: 1967 in S. Matsushita and W. Campbell (eds.), *Physics of Geomagnetic Phenomena*, Academic Press, Vol. 1, Ch. 3.

14.17. Polar Cap Absorption (PCA)

The ionospheric disturbance known as PCA is the strong absorption of radio waves or cosmic radio noise (1–50 MHz) of galactic or extragalactic origin observed in high-latitude polar regions of the Earth. PCA's begin within a few hours of the occurrence of a major solar flare. The actual delay time varies from event to event. The maximum of the PCA event is observed within a day or two of the onset, the recovery may take 10 days or so.

The PCA events are caused by enhanced ionization of the atmosphere between 50–90 km produced by solar 5–20 MeV protons emitted during solar flares. This has been firmly established by rocket and satellite experiments. Since the low-energy protons reach the Earth's atmosphere only in the polar cap region, the absorption of cosmic noise occurs only there, and hence the name PCA.

Little and Leinbach (1959) specially designed receivers called *riometers* (relative ionospheric opacity meter) for the continuous monitoring of cosmic noise in the polar regions in order to observe PCA events. The technique has been found very useful in monitoring low-energy cosmic ray protons around the Earth. The absorption of cosmic noise is very sensitive to the solar particle flux and can give an indirect measure of this flux from the riometer recordings in the polar cap.

Bailey, D. K.: 1964, *Planet. Space Sci.* **12**, 495.
Bryant, D. A., Cline, T. L., Desai, U. D., and McDonald, F. B.: 1962, *J. Geophys. Res.* **67**, 4983.
Little, C. G. and Leinbach, H.: 1959, *Proc. IRE* **47**, 315.
Reid, G. C.: 1967 in S. Matsushita and W. Campbell (eds.), *Physics of Geomagnetic Phenomena*, Academic Press, New York, Ch. 4.

14.18. Sudden Ionospheric Disturbance (SID)

The term sudden ionospheric disturbance (SID) is used for solar-terrestrial events during which increased attenuation of radio waves or cosmic noise is observed. The phenomenon was discovered in short-wave radio signal attenuation in radio communication and is also known as *Mögel-Dellinger effect*. The SID events have an onset of a few minutes and last about an hour. In earlier literature they are referred to as *SWF* (*short-wave fade out*), *SSWF* (*sudden SWF*) and *GSWF* (*gradual SWF*). The observed signal attenuation is caused by increased ionization in the *D* region of the ionosphere. Similar attenuation occurs for the cosmic noise of extraterrestrial origin passing through the *D* region. The short-lived, increased ionization of the *D* region affects also low-frequency and very low-frequency (VLF) propagation producing *sudden phase anomalies* (*SPA*). Moreover, the increased ionization gives rise to enhanced reflection of low-frequency waves generated in the atmosphere by thunderstorms ('atmospherics'); this is known as *sudden enhancement of atmospherics* (*SEA*).

The SID's are solar-terrestrial phenomena associated with the occurrence of solar flares. Flare X-rays $\lambda < 10$ Å are mainly responsible for the enhanced ionization in the *D* region; smaller effects in the *E* and *F* regions are produced by X-rays $\lambda > 10$ Å.

Bauer, S. J.: 1973, *Physics of Planetary Ionospheres*, Springer Verlag, New York, p. 180.
Reid, G. C.: 1967, in S. Matsushita and W. Campbell (eds.), *Physics of Geomagnetic Phenomena*, Academic Press, Ch. 4.

14.19. E Region

The ionosphere between 85–90 and 120–140 km altitudes is named the *E* region (Figure 14.4). *E* region characteristics i.e. electron density, height, etc. depend on the solar zenith angle χ and solar activity (*R* sunspot number):

$$f_0E(\chi, R) = 3.3\,[(1 + 0.008R)\cos\chi]^{1/4}\ \text{MHz},$$
$$N_mE(\chi, R) = 1.35 \times 10^5\,[(1 + 0.008R)\ \cos\chi]^{1/2}\ \text{electrons cm}^{-3}.$$

There is no significant difference in the characteristics of the *E* layer at middle and low latitudes but the low-latitude *E* layer has irregularities arising from the equatorial electrojet. The ionization in the *E* layer is mainly caused by X-rays in the range 8–104 Å. Along with electrons there are positive ions in the *E* region. The major components are O^+ and NO^+.

The *night-time E region* is quite different from the day-time *E* and is considered by some authors to be an independent phenomenon. The electron densities in the night-time *E* layer can be as low as 200 or as high as 10^4 electrons cm^{-3}. The maximum frequency is called f_0IL and maximum density N_mIL. Both day-time and night-time *E* regions have been extensively investigated by rocket-borne Langmuir probes. Radar experiments are being used to study the electric field and the drift velocities of electrons in the *E* region.

Bauer, S. J.: 1973, *Physics of Planetary Ionospheres*, Springer Verlag, New York, Ch. 9.
Morse, F. A. and Rice, C. J.: 1976, *J. Geophys. Res.* **81**, 2795.
Risbeth, H. and Garriot, O. K.: 1969, *Introduction to Ionospheric Physics*, Academic Press, New York, Ch. 5.
Van Zandt, T. E.: 1967, in S. Matsushita and W. Campbell (eds.), *Physics of Geomagnetic Phenomena*, Academic Press, Ch. 3.

14.20. Sporadic *E* (*Es*)

The sporadic *E* is considered to be a phenomenon independent of the normal *E* layer of the ionosphere. The sporadic *E* reflections are due to electron inhomogeneities (layers or patches) on a scale of tens or hundreds of km at 100–120 km height in the *E* region. The IGY definition of *Es* reflection is based on one or more characteristics: (1) random time of occurrence; (2) partial transparency (echoes from higher levels); (3) variations of penetration frequency with transmitter power; (4) virtual height independent of frequency. There are several classifications of *Es* based on the shapes of the traces on ionograms (for details see Smith and Matsushita, 1962).

In high latitudes, the *Es* is associated with aurora, night-time *E* (sometimes also called auroral *Es*) and particle precipitation originating from the tail of the magnetosphere or possibly of interplanetary origin. At middle latitudes, *Es* layers have been explained by the theory of windshear (Whitehead, 1961). The theory is controversial because it fails to explain the quantitative observation of maximum electron density $N_m E_s$ which may be $\approx 10^6$ cm^{-3}. Some of the difficulties in the theory may be overcome if the role of metallic ions (at present poorly known) is considered.

At the equator, the *Es* is associated with the equatorial electrojet ($\approx \pm 5\frac{1}{2}°$ dip angle) and is called the ***equatorial Es*** (*Es-q*) and ***equatorial slant sporadic E*** (*Es-s*). The equatorial *Es* (*Es-q*) is also called Type 2 irregularity (see 14.21). The *Es-q* is not really sporadic in occurrence, but it appears regularly (sometimes 70% of the daytime) at the equator and occasionally suddenly disappears for a period of minutes or hours. Such sudden disappearances, sometimes called *SDEsq*, are possibly related to interplanetary magnetic field changes. The *SDEsq* is sometimes associated with a decrease in the geomagnetic field at the observatory just below the equatorial electrojet. Such a decrease in the *H* component of the geomagnetic field is attributed to a reversal of the electrojet currents called the *counter-electrojet.*

Bauer, S. J.: 1973, *Physics of Planetary Ionospheres*, Springer Verlag, New York, Ch. 9.
Carter, D. A., Balsley, B. B., and Ecklund, W. L.: 1976, *J. Geophys. Res.* **81**, 2786.
Fambitakoye, O., Rastogi, R. G., Tabbagh, J., and Vila, P.: 1973, *J. Atmos. Terr. Phys.* **35**, 1119.
Rastogi, R. G. and Patel, V. L.: 1975, *Proc. Ind. Acad. Sci.* **82A**, 121.
Smith, E. K. and Matsushita, S. (eds.): 1962, *Ionospheric Sporadic E*, The MacMillan Co., New York.

14.21. Equatorial Electrojet

The quiet-day variation of the geomagnetic field is very large ($\approx$ 200 nT) at the dip equator compared to variations of about 20 nT at middle latitudes. This extremely large value is due to a thin electric current layer in the ionosphere over the dip equator which is called the Equatorial Electrojet. This layer is just a few degrees wide (±5° around dip equator) and is at about 100–115 km height as recently verified by rocket experiments. The existence of the electrojet current is the consequence of the highly anisotropic conductivity of the ionospheric plasma between 90 and 130 km heights.

The direct conductivity (σ_0) along the field is quite high so that the field lines are almost electric equipotentials. But the east-west (Cowling) conductivity $\sigma_3 = \sigma_1 + \sigma_2^2/\sigma_1$, where σ_1 is the Pedersen and σ_2 is the Hall conductivity, is comparable to the direct conductivity. The ionospheric conductivity has the following profile:

(1) $\approx$ 70 km $\sigma_0 > \sigma_1 \gg \sigma_2$;
(2) $\approx$ 90 to 130 km $\sigma_0 > \sigma_2 > \sigma_1$; σ_3 very important;
(3) $\approx$ 130 to 160 km $\sigma_0 \gg \sigma_2, \sigma_1$.

The horizontal component of the dynamo force is almost zero along the dip equator. However, the east-west electrostatic field drives an eastward Pedersen current along the dip equator and an associated downward Hall current. Because the conducting layer is a thin shell this downward current creates an upwardly directed polarization electric field whose Pedersen current cancels the downward Hall current. This secondary electric field drives an intense Hall current in the eastward direction strengthening the eastward Pedersen current of the main electric field. This current is the electrojet and is due mainly to westward electron motion in the daytime.

The role of plasma instabilities and inhomogeneities is the subject of current research. There are two types of irregularities with different wavenumber k dependence. *Type 1* has a flat k spectrum at predominantly short wavelengths and has been explained by the two-stream instability which involves ion inertia. It usually appears when the electron drift velocity exceeds the ion acoustic velocity. The *Type 2 irregularity* is excited at long wavelengths ($\geqslant 100$ m) and is the result of a gradient instability or cross-field instability or Simon-Hoh instability (see also 14.20).

Cahill, L. J., Jr.: 1959, *J. Geophys. Res.* **64**, 489.
Farley, D. T.: 1974, *Rev. Geophys. Space Phys.* **12**, 285.
Onwumechilli, A.: 1967, in S. Matsushita and W. H. Campbell (eds.), *Physics of Geomagnetic Phenomena*, Academic Press, New York, Ch. 3.
Prakash, S., Subbaraya, B. H., and Gupta, S. P.: 1971, *J. Atmos. Terr. Phys.* **33**, 129.
Rastogi, R. G. and Patel, V. L.: 1975, *Proc. Ind. Acad. Sci.* **82A**, 121.

14.22. Solar Flare Effect (Sfe)

The solar flare-associated short-wave electromagnetic radiation (i.e. UV, EUV and X-ray) causes a significant and detectable effect on the geomagnetic field. This radiation enhances the ionospheric conductivity and thus temporarily alters the normal ionospheric *Sq* currents in the sunlit side of the Earth. The resultant effect on the geomagnetic field is called an Sfe (solar flare effect) and appears in the form of a kink or impulsive change lasting about 30 min in the H trace of magnetograms. The effect is also called a *crochet* or *Sqa* (*Sq augmentation*). The Sfe magnitude in H may be as large as 50 nT. The Sfe magnetic variation may be positive or negative in the H component depending on the local time of the occurrence and the latitude of the magnetic observatory. Usually the Sfe effect is larger at low latitudes than at middle latitudes. However, contrasting observations have been obtained, presumably due to the concurrent effect of the electrojet and counter-electrojet. The Sfe effects are the opposite of the *solar eclipse effects* when the solar radiation is momentarily cut off.

Dellinger, J. H.: 1937, *Terr. Magn. Atmos. Elect.* **42**, 49.
Richmond, A. D. and Venkateswaran, S. V.: 1971, *Radio Sci.* **6**, 139.
Van Sabben, D.: 1968, *J. Atmos. Terr. Phys.* **18**, 192.

14.23. F_1 and F_2 Regions

The F_1 region of the ionosphere is approximately between 140–230 km altitude (Figure 14.4). The height varies with solar activity, season and geomagnetic activity. Rocket experiments indicate a maximum-density height between 160–180 km, ground ionograms indicate ≈ 200 km. f_0F_1 and N_mF_1 show a dependence on solar zenith angle χ and sunspot number R similar to the E layer dependence:

$$f_0F_1(\chi, R) = 4.25\,[(1 + 0.015R)\ \cos\chi]^{1/4}\ \text{MHz},$$
$$N_mF_1(\chi, R) = 2.25 \times 10^5\,[(1 + 0.015R)\cos\chi]^{1/2}\ \text{cm}^{-3}.$$

The main ionizing solar radiation for F_1 is $\lambda \approx 304$ Å.

The F_2 region has a maximum electron density N_mF_2 between 10^4 and 8×10^6 cm^{-3} and f_0F_2 lies between 1 and 25 MHz. The maximum-density height h_mF_2 is from 200 to 600 km. The F_2 layer is highly variable and shows effects correlated with sunspots, seasons and latitude. At sunspot minimum, the average h_mF_2 at middle latitudes is 225 km in the daytime and 300 km in the night-time in winter. In the summer and at maximum sunspot activity, these values are increased by 25 to 50 km. In contrast to the f_0E and f_0F_1, there is no solar zenith angle dependence of the diurnal and latitudinal variations of f_0F_2. The Sun does affect the electron density of the F_2 layers which shows a rapid increase after sunrise, but, surprizingly, the maximum f_0F_2 may occur on occasions after sunset. The F_2 layer is always present during night-time.

The diurnal, seasonal and sunspot effects are complicated and sometimes confusing. For instance, the f_0F_2 values are systematically larger all over the world in November, December and January than in May, June and July. This effect is known as the *December anomaly*. On the other hand, in high and middle latitudes the daytime f_0F_2 and N_mF_2 values are two and four times larger, respectively, in winter than in summer; this is the *winter anomaly*. Note that the two anomalies are in phase on the northern hemisphere but separated by half a year on the southern hemisphere.

The high variability of the F_2 layer and the possible interaction with the plasmasphere make a definitive explanation of the various effects very difficult.

The F region electron densities are highly variable and show the well-known phenomena of spread F and Travelling Disturbances.

Dyson, P. L., McClure, J. P., and Hanson, W. B.: 1974, *J. Geophys. Res.* **79**, 1497.
Evans, J. V.: 1975, *Rev. Geophys. Space Phys.* **13**, 887.
Van Zandt, T. E.: 1967, in S. Matsushita and W. H. Campbell (eds.), *Physics of Geomagnetic Phenomena*, Academic Press, New York, Ch. 3.

14.24. Spread F

The inhomogeneities in the F region ionospheric plasma density give rise to the phenomenon called spread F. Usually the F region reflections of electromagnetic waves are clear echoes indicating a well-defined plasma layer. But on certain occasions the reflected waves (echoes) are fuzzy indicating that overlapping echoes are coming from different places, i.e. the reflected waves received by the sounder are due to spatial plasma density inhomogeneities from 250 km to the height of N_mF. The size of the irregularities may vary from 20 to more than 100 km. In observations, terms like *range-spreading* (i.e. complete F region echoes displaced in range appear in ionograms) or *frequency-spreading* (i.e. the reflection echo pattern from the F region is spread over a range of frequencies) are used.

The small-scale (less than 40 km) irregularities in plasma density cause diffraction of electromagnetic wave signals originating from stars or satellites; this effect is called *scintillation*. The terms *VHF* and *GHF scintillation* have been used for the appropriate frequency ranges.

Spread F is observed mostly near the equator ±20° geom. lat. and at high latitudes, greater than ±40° geom. lat. It is rarely observed inbetween. It occurs usually near midnight, before midnight in the equatorial zone and after midnight in the high-latitude zone. The spread F at high latitudes is correlated with geomagnetic activity; at the equator, it has no geomagnetic control. Recent observations of Fe^+ by a satellite experiment indicate the importance of such ions in the equatorial spread F.

So far, there is no theory which completely explains the spread F.

Chaturvedi, P. and Kaw, P. K.: 1976, *J. Geophys. Res.* **81**, 3257.
Dyson, P. L., McClure, J. P., and Hanson, W. B.: 1974, *J. Geophys. Res.* **79**, 1497.
Hanson, W. B., McClure, J. P., and Sterling, D. L.: 1973. *J. Geophys. Res.* **78**, 1973.
Herman, J. R.: 1966, *Rev. Geophys.* **4**, 255.

14.25. Travelling Ionospheric Disturbances (TID)

The TID is an irregularity of the F region which has large horizontal dimensions compared to other irregularities. It is a wave-like oscillation of the contours of constant electron density moving slowly downward with time. The disturbance usually travels in unchanged form over a horizontal distance ≈ 1000 km. The travel speed of TID's is ≈ 200 m s^{-1} to ≈ 10 km s^{-1}. The entire disturbance may last for 10–60 min. Observations in the equatorial region suggest that about two thirds of the TID's have horizontal wavelengths of 10 to 100 km and periods of 10–30 min with a phase velocity of 100–200 m s^{-1}. The TID occurrence and direction of travel have diurnal, seasonal and sunspot cycle dependencies. The mechanism generating TID's is not definitely known, but the suggestion has been made that internal gravity waves travelling up to the F region from the lower atmosphere play an important role. Because gravity waves originate from the weather system of the stratosphere and troposphere, a connexion with weather disturbances has been suggested.

Evans, J. V.: 1975, *Rev. Geophys. Space Phys.* **13**, 887.
Georges, T. M.: 1968, *J. Atmos. Terr. Phys.* **30**, 735.
Munro, G. H.: 1950, *Proc. Roy. Soc. London* **A202**, 208.
Röttger, J.: 1976, *Kleinheubacher Berichte* **19**, 443.

14.26. Magnetosphere

The solar wind plasma flowing continuously past the Earth creates a region around the Earth which confines the geomagnetic field. This region which contains plasma and the Earth's magnetic field is known as the magnetosphere. On the average the sunward boundary of the magnetosphere is between 12–10 R_E. The outer boundary layer is called the *magnetopause* and may have a thickness of 100–200 km. Its location depends on the momentum flux density of the solar wind plasma and the value of the magnetic field at that distance and may be found as close as 8–6.6 R_E occasionally. On the night side, the magnetosphere extends like a comet-tail. This region is called the geomagnetic tail or magnetospheric tail or magnetotail (see 14.30); it can be traced as far as 1000 R_E beyond which the distinction between interplanetary and geomagnetic field becomes impossible. (Figure 14.5 and 14.7).

The region surrounding the magnetopause contains post-shock solar wind plasma and is called the magnetosheath. The magnetosheath is bounded by a standing *bow-shock*

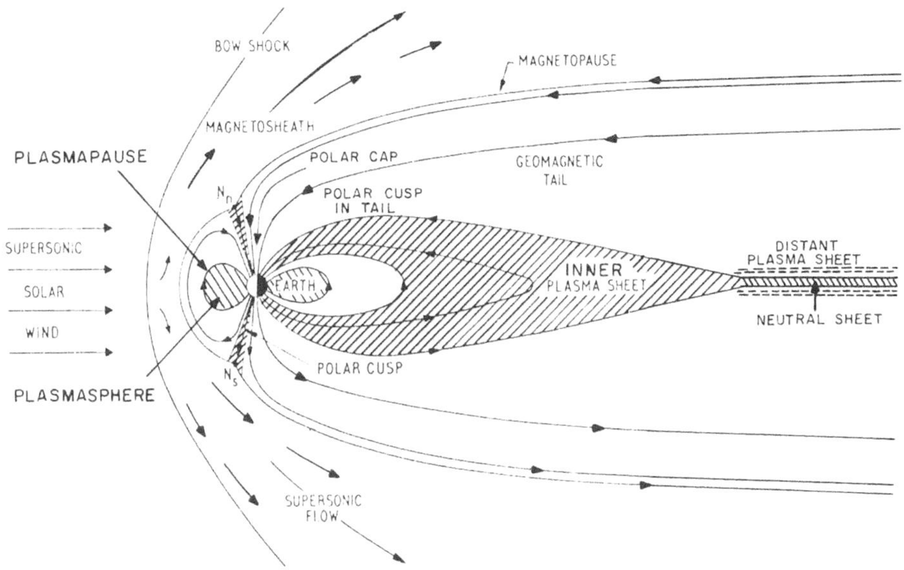

Fig. 14.5. The magnetosphere of the earth and the bow-shock caused by the super-Alfvénic and supersonic solar wind. The subregions of the magnetosphere and their boundary layers are also labelled.

created by the interaction of the super-Alfvénic flow of the solar wind plasma with the magnetosphere. This collisionless hydromagnetic shock-wave has been the subject of intensive study of its internal structure, wave generation, location and movements because such phenomena are difficult to duplicate in laboratory plasma experiments.

The high-latitude geomagnetic field and the corresponding magnetospheric region is complicated. The points N_n and N_s are neutral points around which a funnel-like region of weak magnetic field and plasma with magnetosheath characteristics exist. This magnetospheric region and its polar ionospheric counterpart is called the *polar cusp, cusp, dayside cusp* or dayside *magnetospheric cleft.* The sunward-side edge is defined by the closed field lines that coincide with the dayside magnetopause and the other edge of the cusp boundary is defined by the inner side of the polar cap. The approximate thickness of the dayside polar cusp is 2° to 5° or ≈ 1200 km at the top and 12 km at the surface of the Earth. There is another region in the nightside which is called the *(polar) cusp in magnetotail*. It is defined by the closed field lines (trapping boundary) extending into the plasma sheet in the night magnetosphere on one side, and by open field lines going into the upper edge of the geomagnetic tail on the other side. This cusp defines the night-time boundary of the auroral oval. Between these two cusps is the region which is called the *polar cap.* This region contains magnetic field lines that are vertical and extend into the diverging region of the polar magnetosphere. The field lines are then swept back into the tail region. A similar configuration occurs in the southern hemisphere. The polar cap region open field lines allow a flow of plasma from the ionosphere to the distant tail region which is called the *polar wind.*

Cahill, L. J., Jr. and Patel, V. L.: 1967, *Planet. Space Sci.* **15**, 997.
Fairfield, D. H.: 1971, *J. Geophys. Res.* **76**, 6700.
Formisano, V., Russell, C. T., Means, J. D., Greenstadt, E. W., Scarf, F. L., and Neugebauer, M.: (see p. 240).: 1975, *J. Geophys. Res.* **80**, 2013.
Frank, L. A.: 1975, *Rev. Geophys. Space Phys.* **13**, 974.
Gold, T.: 1959, *J. Geophys. Res.* **64**, 1219.
Willis, D. M.: 1971, *Rev. Geophys. Space Phys.* **9**, 953.

14.27. Magnetospheric Coordinate Systems

In solar-terrestrial studies three geocentric magnetospheric coordinate systems are widely used. The *solar ecliptic coordinate system* (Figure 14.6a) is formed by an X_{SE} axis towards the Sun and a Z_{SE} axis perpendicular (north) to the ecliptic plane. The Y_{SE} axis completes a right-handed coordinate system. It is referred to simply as *SE coordinates.*

In the *solar magnetospheric system* (Figure 14.6b) the X_{SM} axis is the same as the X_{SE} but the Z_{SM} axis lies in the plane formed by the X_{SM} axis and the Earth's dipole axis; the Y_{SM} axis completes the right-handed system. This system is referred to as *GSM* or *SM.*

The third coordinate system is the *solar magnetic system.* In these coordinates the Y_{MG} axis is the same as the Y_{SM} axis and the Z_{MG} axis is coincident with the dipole axis. The X_{MG} axis lies in the plane formed by the Sun-Earth line and the dipole axis (Figure 14.6c).

For studies of trapped particles in the geomagnetic field, *McIlwain's coordinates* or *B-L coordinates* were introduced. This system depends on the magnitude of the magnetic field B and the integral invariant I defined as

$$I = \int_A^{A'} \left(1 - \frac{B_l}{B}\right)^{1/2} ds.$$

The integral is taken along the line of force between two conjugate points. B is the magnetic field strength at point A, and B_l that along the line of force. McIlwain defined a parameter, called *McIlwain's L parameter* which is a function of I and B. For a dipole

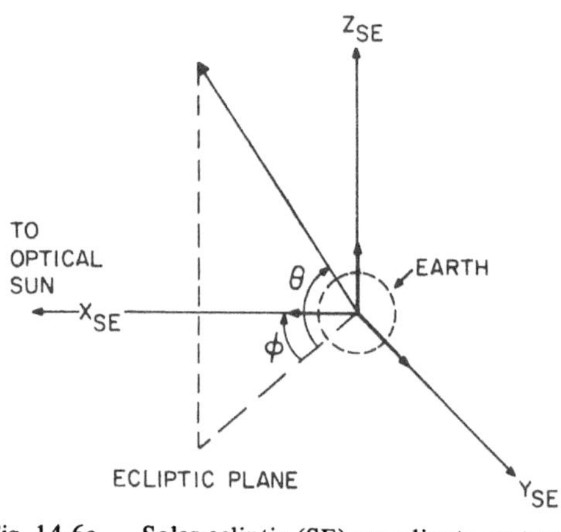

Fig. 14.6a. Solar-ecliptic (SE) coordinate system.

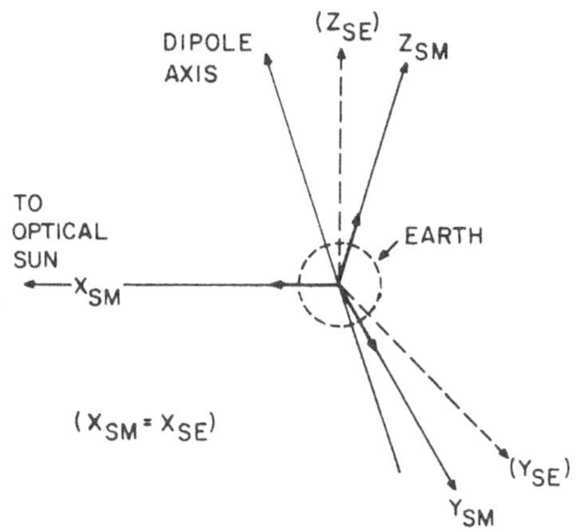

Fig. 14.6b. Solar-magnetospheric (SM) or Geocentric solar-magnetospheric (GSM) coordinate system.

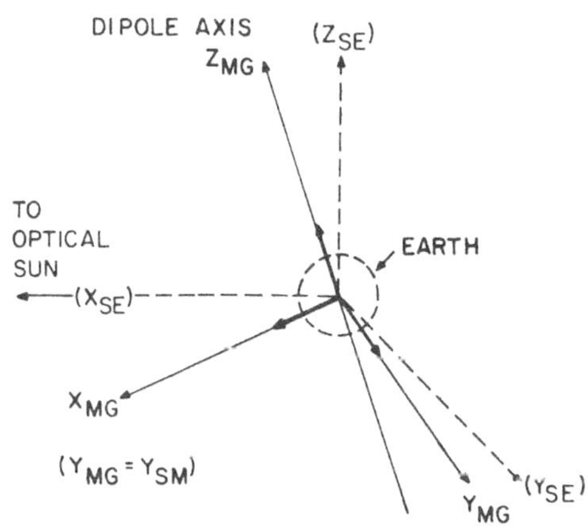

Fig. 14.6c. Solar-magnetic (MG) coordinate system.

field, McIlwain's L value is the radial distance (in R_E) in the equatorial plane to the points of intersection of field lines of given B.

In geomagnetism and solar-terrestrial studies, *polar coordinates* R, Λ are also used which can be calculated from the B, L coordinates. Λ is called the *invariant latitude* (for $R = 1$) and is obtained from $\cos^2 \Lambda = 1/L$.

McIlwain, C. E.: 1961, *J. Geophys. Res.* **66**, 3681.
Ness, N. F.: 1965, *J. Geophys. Res.* **70**, 2989.
Roederer, J. G.: 1970, *Dynamics of Geomagnetically Trapped Radiation*, Springer-Verlag, New York, Ch. 4.

14.28. Plasmasphere

Above the ionosphere (at altitudes $\gtrsim 1000$ km), there is a region surrounding the Earth which contains high density cold plasma. This region is called the plasmasphere; it extends up to 3 R_E, and sometimes as far as 7 R_E. The plasma density slowly decreases with height until it abruptly drops by a factor of 100 at the boundary of the plasmasphere. This boundary is called the *plasmapause*. It has a thickness of a fraction of one Earth radius. Typical plasma density changes at the plasmapause near 1200 local time are from 10^3 ions cm^{-3} to 10 ions cm^{-3}. The plasma in the plasmasphere consists of up to 99% protons and electrons with small fractions of He^+ and O^+. The plasmasphere is not spherical but has a bulge in the region between 1500 and 2200 local time. This is called the *plasmasphere bulge* and may extend up to 2 to 3 R_E farther than the average dimension of the region. The plasma density in the bulge varies as R^{-4}. Solar-terrestrial studies show that the plasmasphere becomes smaller with increasing geomagnetic activity, i.e. with higher *Kp* index. (Figures 14.5 and 14.7).

Carpenter, D. L.: 1963, *J. Geophys. Res.* **68**, 1675.
Carpenter, D. L. and Park, C. G.: 1973, *Rev. Geophys. Space Phys.* **11**, 133.
Chappell, C. R.: 1972, *Rev. Geophys. Space Phys.* **10**, 951.
Mozer, F. S.: 1973, *Rev. Geophys. Space Phys.* **11**, 755.
Nishida, A.: 1971, *J. Geophys. Res.* **71**, 5669.

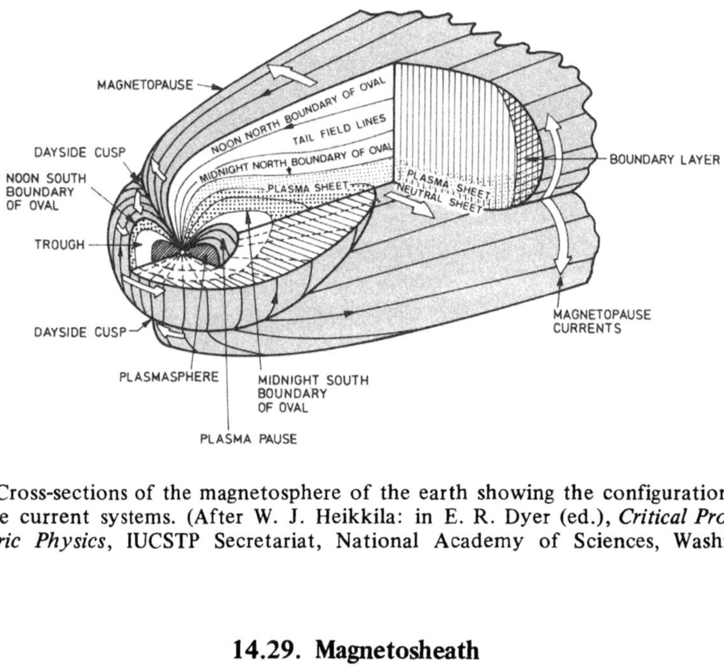

Fig. 14.7. Cross-sections of the magnetosphere of the earth showing the configuration of the oval, cusps and the current systems. (After W. J. Heikkila: in E. R. Dyer (ed.), *Critical Problems of the Magnetospheric Physics*, IUCSTP Secretariat, National Academy of Sciences, Washington, D.C., U.S.A.)

14.29. Magnetosheath

The region between the magnetopause and the standing bow-shock is called the magnetosheath (Figure 14.5). It is characterized by post-shock thermalized solar wind plasma and a turbulent magnetic field. The plasma flow speed (≈ 250 km s^{-1}) is considerably reduced compared to interplanetary values and the direction of the solar wind plasma is deflected (as much as 20°). The plasma electron and proton temperatures are increased and are typically $\approx 10^6$ K. The magnetosheath near the Sun-Earth line is approximately between 10 and 14 R_E, and the region flares out away from this line. In the dawn-dusk region the magnetosheath on the average extends from 14 to 22 R_E.

Fairfield, D. H.: 1976, *Rev. Geophys. Space Phys.* **14**, 117.
Howe, H. C. and Binsack, J. H.: 1972, *J. Geophys. Res.* **77**, 3334.

14.30. Magnetospheric Tail

A portion of the Earth's magnetosphere begins near 8 to 10 R_E on the nightside and extends like a comet-tail in the antisolar direction up to 80 and possibly as far as 1000 R_E; it is referred to as the *geomagnetic tail, geomagnetotail, magnetotail* or *magnetic tail.*

In the near-Earth central region of the nightside tail from approximately 8 to 30 R_E there is a reservoir of energetic plasma particles. This region is called the *inner plasma sheet* (Figures 14.5, 14.7). The geomagnetic field lines are still closed but extended into a tear-shaped configuration. The dynamics of this region plays an important role in high-latitude geomagnetic phenomena such as substorms. The distant geomagnetic tail ($\gtrsim$ 30–40 R_E) has in the central part a *neutral sheet* which arises from the merging of the field lines in the lower part of the tail directed in the antisolar direction with the field lines in the upper part of the tail directed towards the Sun. Around this neutral sheet is a region of 4 to 6 R_E thickness which contains hot plasma (Figure 14.5). This region is called the *plasma sheet* or *distant plasma sheet.* It contains plasma at densities of 0.1 to 1 particles cm^{-3}. The temperature of the electrons is $\approx$ 100–500 eV and that of the protons is $\approx$ 1–5 keV.

The geometry of the distant tail may be approximated by two cylinders of approximate radius 20 R_E. The upper cylinder has magnetic field lines along the sunward direction and the lower one field lines in the antisolar direction. The magnetic field slowly decreases in the tail region from 20 to 30 nT around 20 R_E to 5 to 10 nT at distances greater than 60 R_E. The orientation of the tail depends mostly on the solar wind direction. However, the direction of the interplanetary field (especially the Z component in the ecliptic plane) affects the dynamics of the tail and associated solar-terrestrial phenomena at high-latitude regions of the Earth such as aurora, substorms, etc. The approximate uniformity of the magnetic field at large distances and the clearly defined geometry allow the study of linear and non-linear waves in the tail which is a natural laboratory for low-density plasma studies. It has been suggested that the tail region is the possible source of long-period micropulsations.

Hill, T. W.: 1974, *Rev. Geophys. Space Phys.* **12**, 379.
Ness, N. F.: 1965, *J. Geophys. Res.* **70**, 2989.
Patel, V. L.: 1968, *Nature* **218**, 857; *Phys. Letters* **26A**, 596.
Russell, C. T. and McPherron, R. L.: 1973, *Space Sci. Rev.* **15**, 205.
Schindler, K.: 1975, *Space Sci. Rev.* **17**, 589.

14.31. Substorm

In solar-terrestrial studies, geomagnetic perturbations of duration one to two hours occurring near midnight local time have been termed *polar magnetic substorms* or simply substorms. The perturbations are not part of the phenomenon called a storm; they occur independently and have entirely different physical causes. A typical substorm is a magnetic perturbation of several hundred nT magnitude in the auroral zone (Figure 14.8). Substorms may occur in a series, as composite substorms lasting 2 to 3 hr.

At lower latitudes, the corresponding magnetic perturbation has the appearance of a bay on the coast of an ocean. The term *geomagnetic bay* has been used since the nineteenth century for negative or positive changes in *H, D* and *Z* components of the geomagnetic field.

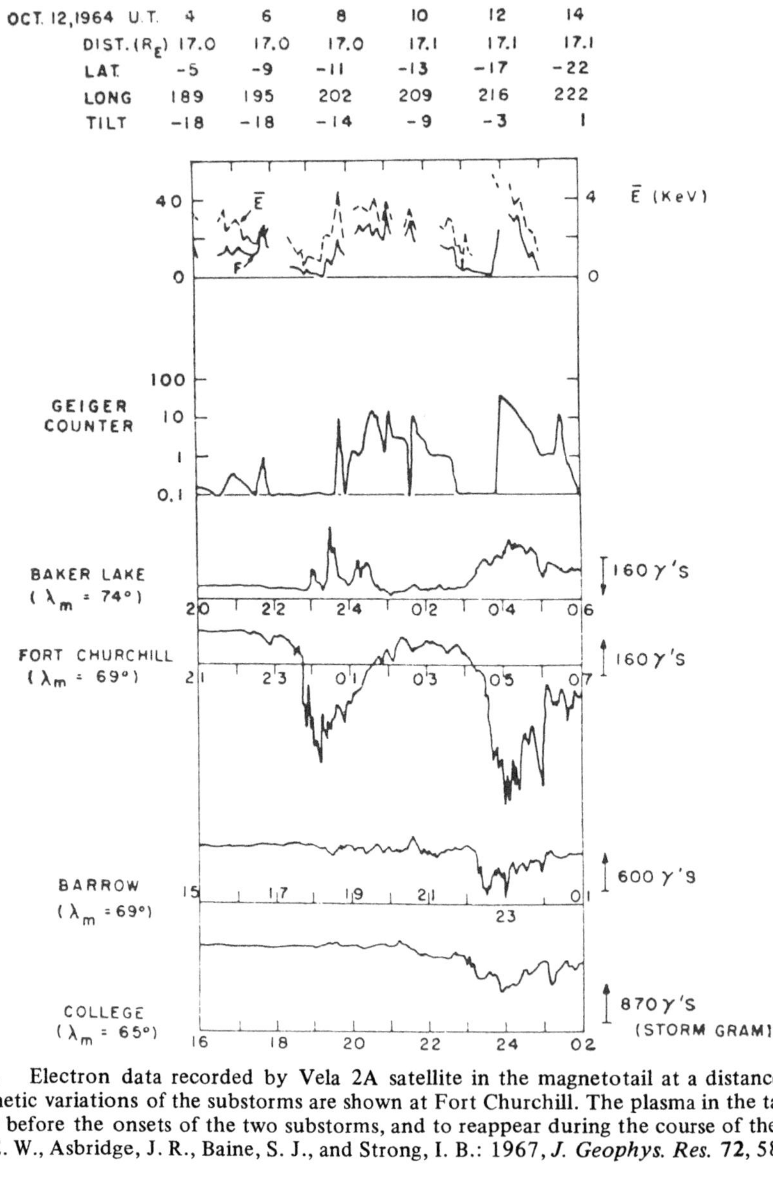

Fig. 14.8. Electron data recorded by Vela 2A satellite in the magnetotail at a distance of 17 R_E. The magnetic variations of the substorms are shown at Fort Churchill. The plasma in the tail is seen to disappear before the onsets of the two substorms, and to reappear during the course of the substorms. (Hones, E. W., Asbridge, J. R., Baine, S. J., and Strong, I. B.: 1967, *J. Geophys. Res.* **72**, 5879.)

The substorm-associated magnetic perturbations have been interpreted in terms of ionospheric current systems (*Birkeland currents*); these are modified by field-aligned currents from the magnetosphere that enter the polar ionosphere intensifying the westward auroral electrojet currents. It should be noted that field-aligned currents of $\approx 3 \times 10^7$ A m^{-2} may exist in quiet times near the auroral oval.

The corresponding perturbations in the magnetosphere associated with polar substorms are called *magnetospheric substorms.* Actually, magnetospheric substorms which are initiated by changes in the interplanetary field orientation and interaction with the geomagnetic field are considered to be the basic phenomenon. The magnetic field in the tail region builds up before the onset of a substorm and relaxes to the normal configuration during the substorm. Apparently energy is fed into the magnetosphere during the build-up process and is dissipated when the tail field returns to normal conditions. The associated disturbances in the ionosphere are called *ionospheric* substorms.

An *auroral substorm* is defined as an aurorally active period of ≈ 1 hr near magnetic midnight. It begins with quiet arc and band auroras. The onset of an auroral substorm is marked by the brightening of the bands followed by the deformation and poleward movement of the arcs near magnetic midnight meridian. The ray auroras are formed in place of the arcs. The maximum activity phase consists of rapid motions and intensity variations all along the auroral oval. The original bands and new arcs move away from the initial positions. The end of the maximum activity phase is marked by the appearance of weak and diffused forms of auroras and substorms end with the reappearance of the arcs and bands. Usually 2 or 3 hr elapse before another auroral substorm takes place.

There is another type of classification for polar magnetic variations labelled *DP* by Nishida. In this classification *DP1* corresponds to polar magnetic substorms. Its equivalent current system has four vortices all in contact with the auroral oval. The *DP2* current has twin vortices which lie in the morning and evening meridians and extend from the polar cap to low latitudes. The *DP2* variations are more responsive to changes in the B_z interplanetary field component and are assumed to be connected with magnetospheric convective motion.

Anderson, H. R. and Vondrax, R. R.: 1975, *Rev. Geophys. Space Phys.* **13**, 243.
Arnoldy, R. L.: 1974, *Rev. Geophys. Space Phys.* **12**, 217.
Barcus, J. R.: 1972, *Space Sci. Rev.* **13**, 295.
Cloutier, P. A. and Anderson, H. R.: 1975, *Space Sci. Rev.* **17**, 563.
Iijima, T. and Potemra, T. A.: 1976, *J. Geophys. Res.* **81**, 2165.
Nishida, A.: 1971, *Cosmic Electrodynamics* **2**, 350.

14.32. Radiation Belts or Van Allen Belts

The motion of charged particles in a dipolar geomagnetic field is such that particles of a certain energy can be trapped. Actually, there are two Van Allen belts or radiation belts around the Earth: the *inner radiation belt* is located between 1.2 and 4.5 R_E and contains protons of MeV range which are called inner belt protons. The *outer radiation belt* is between 4.5 and 6.0 R_E and has largely protons with energies between 200 eV and 1 MeV. They are called outer belt protons.

The *inner belt protons* largely originate from cosmic ray albedo neutrons. Cosmic rays entering the Earth's atmosphere have collisions with nitrogen and oxygen atoms resulting in neutrons. These neutrons produced by the knock-on process have energies between 1 MeV and 1 GeV. Another type of neutron produced by excitation of nuclei has energies of 8 MeV or more. These neutrons decay into protons and electrons. Inner belt protons of energy >4 MeV and energy >50 MeV are observed experimentally. The high energy inner belt protons (>50 MeV) have a peak flux $\approx 10^4$ cm^{-2} s^{-1} around 1.5 R_E and at 2.2 R_E. The protons of energy >4 MeV have a peak flux $\approx 10^6$ cm^{-2} s^{-1} near 2.0 R_E. Ideally, in a static dipole, the particles could be trapped permanently. But the protons of energy >1 GeV are lost by interactions with atmospheric atoms and protons of 1 MeV to 1 GeV are mostly lost by coulomb scattering. Protons of energy less than 1 MeV are lost by charge-exchange processes.

The *outer belt protons* of 200 eV to 1 MeV are believed to come from the solar wind by diffusion at the magnetopause. Magnetic and electric field fluctuations (of which quantitative estimates are hard to obtain) must be considered in the diffusion process. Drifting particles trapped on dayside field lines may not find the appropriate field lines in night-time and will be lost. These are called *pseudo-trapped particles.*

Electrons with energies >40 keV are found throughout the trapping region. The upper limit to the flux of these electrons is given by the Kennel-Petschek formula in noon regions

$$F = \frac{7 \times 10^{10}}{L^4} + 10^6 \text{ electrons cm}^{-2}\text{ s}^{-1}.$$

These electrons play an important role in several solar-terrestrial phenomena. During a magnetospheric substorm large numbers of electrons appear in the magnetotail and trapping region. When the pitch angle (the angle between the velocity vector of the particle and the magnetic field vector) distribution becomes anisotropic VLF waves are generated. The term trapped particle or radiation is used not only for Van Allen belt particles but for all particles that can be trapped in the magnetosphere up to the magnetopause.

Particles trapped in the dipolar magnetic field execute three basic motions according to three *adiabatic invariants.* The *first adiabatic invariant* or magnetic moment of the particle is

$$\mu = \frac{\frac{1}{2} m v_\perp^2}{B}.$$

As long as spatial and time variations in the field are such that μ is conserved, the particle gyrates around the field line. The *second adiabatic invariant* holds if the space-time variations are negligible over the bounce period of the particle. Then the particle continues to perform a bouncing motion along the field line between north and south points (mirror points) on the field line. The *third adiabatic invariant* is a flux invariant. The magnetic flux over the surface of a shell is conserved if the time and space variations are negligible over a drift period around the Earth. Fluctuations (e.g. hydromagnetic waves) of appropriate period can violate the three adiabatic invariants and thus remove particles from the trapping regions.

Roederer, J. G.: 1970, *Dynamics of Geomagnetically Trapped Radiation*, Springer-Verlag, New York, Ch. 5.
Schultz, M.: 1975, *Space Sci. Rev.* **17**, 481.
Schultz, M. and Lanzerotti, L. J.: 1974, *Particle Diffusion in the Radiation Belts*, Springer-Verlag, New York, Ch. 2 and 3.
West, H. I., Jr.: 1975, *Rev. Geophys. Space Phys.* **13**, 943.

14.33. Whistlers

The audio frequency electromagnetic waves, frequencies 300–30 000 Hz, initiated in the Earth's atmosphere by lightening are known as *atmospherics.* On entering the ionosphere, they appear as quasi-longitudinal right-hand circularly polarized (QL-R) waves and may be guided back to Earth by the magnetic field. These are whistling atmospherics or whistlers. They have group velocities $v_g \sim \sqrt{f}$, i.e. the group velocity increases with frequency, so that the received signal is a descending tone, hence the name whistler. At middle and low latitudes it is given approximately by

$$v_g = 2c\sqrt{ff_H}/f_p$$

for waves with $f < 10$ kHz, where f_p, f_H are the plasma and the electron gyro frequencies. Whistlers have frequencies $f < f_H$.

The usual whistler observations are displayed on frequency-time diagrams (*sonagrams*). In high latitudes, the group velocity becomes a maximum at frequencies $f_n \approx \frac{1}{4} f_H \cos\theta$ where θ is the angle between the wave normal and the magnetic field. The frequency f_n is called the *nose frequency. Nose-whistlers* occur in groups confirming that whistlers originate from lightning flashes which generate several frequency waves. These travel along the magnetic field lines through the magnetosphere in ducts or columns (sometimes called *ducted whistlers*). Whistler wave ray-directions form a cone of semiangle 19° around the magnetic field directions. Whistlers are reflected from conjugate points in the opposite hemispheres and are observed again after some time delay. The time of propagation is used to calculate the electron density of the magnetosphere. Whistler observations give an excellent opportunity of sounding the plasmasphere and magnetosphere as far as 7 to 9 R_E.

Recent satellite experiments have observed *non-ducted whistlers* in the magnetosphere that do not follow magnetically aligned electron irregularities. The non-ducted waves cannot penetrate the ionosphere and are not observed on the ground.

There are also satellite observations of *ion whistlers* for protons and helium and possibly other heavy ions. They are left-hand polarized waves approaching the local proton or helium ion gyrofrequencies. There is another class of whistlers near the lower hybrid frequency resonance which are called *LHR whistlers.* The mechanism producing whistlers in the magnetosphere is not clear.

Whistler wave packets (solitons) in the solar corona have been proposed in order to explain intermediate-drift bursts in Type IV continua. Coronal whistlers, if present, cannot escape towards the Earth.

Helliwell, R. A.: 1969, *Rev. Geophys.* 7, 281.
Russell, C. T. and McPherron, R. L.: 1972, *Space Sci. Rev.* **12**, 810.
Stix, T. H., 1962: *The Theory of Plasma Waves*, McGraw-Hill, New York, Ch. 2.

14.34. VLF Emissions

There are low-frequency waves of natural origin other than whistlers which are called generally VLF (very low frequency) emissions. In the literature the term VLF is used for frequencies of 3 to 30 000 Hz and *ELF* (extremely low frequency) for 5 to 3000 Hz. The VLF are not emitted by lightning signals but some of them tend to appear with whistlers. Their origin is not completely understood but is related to wave-particle interactions. There are three well-known VLF emissions:

(1) *Periodic VLF* consists of short bursts repeating at regular intervals of a few seconds. Both dispersive and non-dispersive periodic VLF have been observed;

(2) *Hiss* is continuous broad-band emission in the frequency range of 4 to several hundred kHz with a spectral peak near 10–20 kHz. The hiss observed on the Earth usually occurs near geomagnetic midnight and is closely related to auroral activity (homogeneous arcs and bands and auroral break-ups). These VLF emissions are called *auroral hiss*. Auroral hiss has been observed on the ground and in satellites. Occasionally, hiss is observed in the auroral zone without auroral phenomena. *Mid-latitude hiss* is a quite different phenomenon. It occurs with chorus and is not connected with auroras. This term is also used for VLF noise with a steady low-frequency cutoff at mid-latitudes and varying cutoff at high latitudes. This VLF noise has been observed only in satellites in both high and low latitudes.

(3) *Chorus* or *dawn chorus* are VLF emissions usually associated with hiss. They are discrete, well-defined, short-duration tones with rising or falling characteristics or a combination of both. The discrete elements may be repeated many times per second and the resultant sound resembles the waking sound made by birds in the morning. Hence the name dawn chorus or just chorus. The chorus usually occurs with a background hiss or at the upper edge of the hiss band. The chorus observed in the auroral zones is called the *polar chorus* and is usually below 1500 kHz. It is ducted by polar chorus ducts.

Observations of ELF and VLF waves in the magnetosphere have been made by satellite experiments. The theoretical explanation of these phenomena is still the subject of current research.

Parady, B. K., Eberlein, D. D., Marvin, J. A., Taylor, W. W., and Cahill, L. J. Jr.: 1975, *J. Geophys. Res.* **80**, 2183.
Russell, C. T. and McPherron, R. L.: 1972, *Space Sci. Rev.* **12**, 810.

14.35. Aurora

The aurora is a display of beautiful colour patterns of varying intensity and with rapid motions in the high-latitude regions of the Earth. The *visible aurora* contains the green (5577 Å) and red (6300/6364 Å) emission lines from atomic oxygen and molecular bands of N_2 which are excited by energetic particles of solar and magnetospheric origin. These emissions usually occur near or above 100 km height. The term *optical aurora* is used for visible auroras and the auroral emission spectrum from the infrared to the ultraviolet. The energy of the emission in the non-visible spectrum greatly exceeds that in the visible part of the spectrum.

During the appearance of auroras, emission in the VLF ($<$ 30 kHz) frequency range, including VLF chorus and VLF hiss, has been observed. Recently, radio noise emission at 28 MHz and 225 MHz from auroras has been observed. However, the term *radio aurora* is used for the auroral activity creating field-aligned irregularities at auroral heights which produce back-scatter of radio waves.

It should be noted that a large number of solar-terrestrial phenomena such as auroral X-rays, absorption of cosmic noise, radio star scintillation, sporadic E, electric currents in the ionosphere, geomagnetic micropulsations, etc. take place during auroral activity. (For latitude and diurnal dependence see Auroral Oval.)

The occurrence of auroras is correlated with the sunspot cycle, the 27-day cycle, the seasons and magnetic activity. The term *occurrence* is used as a visual auroral index; it is the percentage of hours or nights on which auroral forms are detected, either within a limited part of the sky or in the whole sky. It is not a very good index because it ignores the brightness and extent of the auroral form. A more meaningful index is the *incidence* which measures the number of auroral forms visible overhead in the sky within a given time interval.

Hultqvist, B.: 1967, in S. Matsushita and W. Campbell (eds.), *Physics of Geomagnetic Phenomena*, Academic Press, New York, Ch. 4.
Omholt, A.: 1971, *The Optical Aurora*, Springer-Verlag, New York, Ch. 1.

14.36. Auroral Forms

The actual forms of the aurora are difficult to classify; the most commonly used terms are as follows:

(1) *Quiet homogeneous arcs or bands*: The arc extends about 1000 km in the geomagnetic E-W direction (along the solar direction in polar regions) and has a width of one to several tens of km. The band is the generalization of the arc; it does not have the regular shape of the arc but is usually folded in the form of the letter S or into spirals. They usually occur at 100–150 km heights.

(2) *Auroral rays*: This term refers to an auroral structure elongated along the magnetic field lines with a length of a few tens to several hundred km. The horizontal extent is narrow, a few tens of meters to several kilometers. They usually occur in arcs or bands or as separate structures.

(3) *Patch or surface aurora*: This is an isolated region of luminosity with no particular shape. The patches may be interconnected.

(4) *Veil aurora*: An uncommon form of aurora, it has a uniform luminosity covering large portions of the sky.

The auroral structure is classified as homogeneous, striated or rayed. Several terms are used such as pulsating arc, pulsating surface, diffuse surface, rayed band, drapery, etc.; a review of auroral forms and their classification is given in the International Auroral Atlas (1963).

In the literature, reference is made to auroras classified by colours. In this classification, *Type A auroras* have a red (6300/6364 Å) upper portion or are entirely red in colour. They occur at heights of 300–400 km usually during high geomagnetic activity. *Type B auroras* have lower portions of red colour arising from N_2 First Positive and O_2 First Negative bands. They usually occur during the most active phases of auroral display.

International Auroral Atlas: 1963, Edinburgh University Press.

14.37. Auroral Oval and Auroral Zones

The auroral zones are the zones of maximum night-time occurrence as seen by observers at a fixed point on the Earth; they are located at about 67° north and south latitudes and are about 6° wide. The maximum instantaneous occurrence of auroras in dipole geomagnetic local time occurs in oval-shaped belts (auroral ovals) which are asymmetric around the geomagnetic north and south poles. The auroral oval is fixed on the latitude-time grid and the auroral zone is the locus, on a latitude-longitude grid, of the midnight region of the oval. The oval belt is approximately 23° from the geomagnetic pole during night-time and 15° from the pole in daytime.

The position of the oval belt depends on geomagnetic activity. The oval becomes wide during high geomagnetic activity. The auroral zones or the boundary of the auroral oval are better represented by an L value 6.4 than by dipole coordinates. The dayside oval boundary has field lines coinciding with the magnetopause and the nightside boundary has field lines going into the plasma sheet (Figure 14.7). A variation of the auroral oval with the angle between the geomagnetic axis and the Earth-Sun direction has been observed. The oval has been defined also on the basis of the electron or proton particle precipitation of a particular energy. It may alternatively be defined by the dayside cusp and the cusp in the magnetotail.

The diurnal variation of the occurrence of aurora at the oval shows a maximum at geomagnetic midnight and a minimum at geomagnetic mid-day. On the equatorward side of the oval the auroral occurrence drops sharply but has a similar diurnal variation. On the poleward side of the oval, the occurrence falls gradually and has a complicated diurnal pattern.

Eather, R. H.: 1973, *Rev. Geophys. Space Phys.* **11**, 155.
Frank, L. A.: 1975, *Rev. Geophys. Space Phys.* **13**, 974.
Maynard, N. C.: 1974, *J. Geophys. Res.* **79**, 4620.

14.38. Auroral Intensity

The intensity of auroras is defined by the measurement of the apparent surface brightness. The surface brightness I of an aurora in a particular direction gives the total emission $4\pi I$ in photons cm^{-2} s^{-1}. Because this is not a true surface brightness but emission from a column, the units usually used in auroral research are photons cm^{-2} (column^{-1}) s^{-1}. The usual unit for the total emission is the Rayleigh (R) equal to 10^6 photons cm^{-2} (column^{-1}) s^{-1}. More useful auroral intensity units are defined with reference to the emission in one line or band. For example, the auroral intensity is assigned an *International Brightness Coefficient* (*IBC*) with reference to the green oxygen line (5577 Å); 1 kiloRayleigh (kR) = IBC I, 10 kR = IBC II,

100 kR = IBC III, 1000 kR = IBC IV (maximum intensity of aurora). The classification may not be useful in intense red auroras.

Chamberlain, J. W.: 1961, *Physics of Aurora and Airglow*, Academic Press, New York, Ch. 2.
Davis, T. N.: 1968, in W. N. Hess and G. D. Mead (eds.), *Introduction to Space Science*, Gordon and Breach, New York, Ch. 5.
Omholt, A.: 1971, *The Optical Aurora*, Springer-Verlag, New York, Ch. 1.

14.39. Stable Auroral Red Arcs (SAR Arcs)

Other terms used for SAR arcs are *mid-latitude red arcs* or *M-arcs*. The SAR arc is a subvisual, broad arc that has an east-west extension of thousands of kilometers and possibly extends around the Earth. The latitudinal extension of the arc is 600 km. The SAR is almost monochromatic in the red lines 6300 and 6364 Å. Recently, weak emission lines of 5577 (O I) and 4278 N_2^+ have also been reported. The SAR arcs are classified as auroras but they occur at heights well above the usual height of polar auroras. The lower border is at 300 km, the upper limit is at about 700 km. The intensity of the SAR arc in the 6300 Å emission is between 1 and 10 kR (typical value 6 kR). The visible threshold at this wavelength is about 10 kR so the SAR arcs are rarely visible. However, recent observations have shown that they occur with brightness > 50 kR on 10% of nights. The usual lifetime of the arcs is about one day and they rarely appear on successive days.

Radio waves from satellites or radio sources traversing SAR arcs show scintillation indicating electron density irregularities. The theoretical explanation of SAR arcs is that heated electrons from the ionospheric *F* region are responsible for oxygen atom excitations. Satellites have observed increased electron temperature along geomagnetic field lines that intersect SAR arcs. The SAR arc intensity is positively correlated with geomagnetic activity (storms) and the occurrence of SAR arcs is positively correlated with sunspot activity.

Eather, R. H.: 1975, *Rev. Geophys. Space Phys.* **13**, 925.
Hoch, R. J.: 1973, *Rev. Geophys. Space Phys.* **11**, 935.
Newton, G. P., Walker, J. C. G., and Meijer, P. H. E.: 1974, *J. Geophys. Res.* **79**, 3807.
Omholt, A.: 1971, *The Optical Aurora*, Springer-Verlag, New York, Ch. 1.
Rees, M. H. and Roble, R. G.: 1975, *Rev. Geophys. Space Phys.* **13**, 201.
Russell, C. T. and McPherron, R. L.: 1973, *Space Sci. Rev.* **15**, 205.
Zmuda, A. J., Armstrong, J. C., and Heuring, F. T.: 1970, *J. Geophys. Res.* **75**, 4757.

14.40. Pulsing Aurora

Several auroral forms display almost periodic and coherent time variations of intensity. Those auroras with approximately stationary geometry and rapid periodic in-phase time-variations are called pulsing aurora. They are designated *p-auroras* in the auroral form definition of the International Auroral Atlas. Further subclassifications of *p*-aurora are:

p_1 (*pulsating aurora*) are auroras with uniform phase variation of brightness throughout the form. The ideal definition of pulsating auroras is that stationary space- and time-parts can be separated, i.e. the brightness $I(r,t) = I_s(r)I_T(t)$. Typical p_1-auroras have pulsations of 0.01 to 10 Hz with a low intensity of 1–2 kR. Most p_1-auroras are patches or arcs with a period of several seconds.

p_2(*flaming aurora*): This term is usually used for flame-like surges of luminosity filling the sky rather than for a particular form. They have arc-like geometry and usually move upward from 100 km. These auroras are relatively rare and are more likely to occur in cisauroral than in auroral zones.

p_3(*flickering aurora*) are auroras with rapid, irregular or regular time-variations in brightness giving the appearance of a flickering flame in the sky. They usually occur around the time of auroral break-up. The typical frequency observed in p_3-auroras is 10 ± 3 Hz.

There is another term, *streaming aurora*, used for a class of pulsing aurora which refers to irregular brightness variations moving rapidly horizontally in auroral arcs and bands.

The pulsing aurora is one of the solar-terrestrial phenomena involving pulsations in the geomagnetic field and auroral X-rays produced by precipitating particles of solar and magnetospheric origin.

International Auroral Atlas: 1963, Edinburgh University Press.
Omholt, A.: 1971, *The Optical Aurora*, Springer-Verlag, New York, Ch. 7.
Rosenberg, T. J.,Trefall, H., Kvifte, G. J., Omholt, A., Egeland, A.: 1971. *J. Geophys. Res.* **76**, 122.

14.41. Polar Glow Aurora

The polar glow auroras are characterized by a strong intensity band of First Negative N_2^+ at 3914 Å. Typically, these N_2^+ bands are intensified by a factor of five compared to the (O I) green line at 5577 Å. The absolute intensity of the polar glow aurora is between 0.1 and 10 kR (typically 1–3 kR). These auroras occur with PCA events. A uniform glow is created over the entire polar cap up to 60° geomagnetic latitudes at heights between 30 and 80 km presumably by solar protons and α particles of 10–100 MeV which give maximum ionization at these heights.

There is another glow in the auroral zones called *mantle aurora*. In this type of auroral glow the diurnal maximum intensity is between 1 and 10 kR and occurs in the morning hours; the minimum intensity is 5 times less. There are not many definitive observations of mantle auroras but it appears that the intensity may depend on geomagnetic and solar activity.

Sandford, B. P.: 1961, *Nature* **190**, 245.
Sandford, B. P.: 1967, *Space Res.* **7**, 836.

14.42. Airglow

The airglow is defined as the light produced and emitted by the atmosphere of a planet. It is the non-thermal radiation of the atmosphere excluding auroral emission, lightning and meteor trains. However, the term airglow is usually used for the Earth's atmosphere. In the literature the terms *nightglow, twilightglow* and *dayglow* are used and are self-explanatory. The airglow is only one component of the light of the atmosphere. Other sources are starlight, zodiacal light and daytime scattered light of the Sun. At times the airglow may amount to as much as 40% of the total light.

The airglow originates in atmospheric layers of varying heights and thicknesses. The spectrum of the airglow ranges in wavelength from 1000 Å to 22.5 μ. The major line emission of the airglow is 5577 Å which arises at 90–100 km heights from a layer 30–40 km thick. The emission is due to the *Chapman mechanism* based on the recombination of the oxygen atoms. Other important emission lines are 6300 Å from dissociative recombination of $O_2{}^+$ and emissions of N I 5198/5201 Å and Na I 5890/5896 Å.

The intensity of the airglow is measured in units of Rayleigh. The brightness, in Rayleigh, is $4\pi\beta$ where β is the angular surface brightness of the emitting layer in units of 10^6 photons cm^{-2} sterad^{-1} s^{-1}. The airglow intensity shows varying latitude dependence for the various emission lines. It also shows a diurnal effect with maximum near midnight. The correlation with geomagnetic activity is not clear and contradictory results exist in the literature. However, positive correlations have been reported for 5577 Å airglow with sunspot number and 10.7 cm solar flux. The airglow has been observed by satellite-borne experiments. From outer space it looks like a ring of light around the Earth and is of greenish colour.

Chamberlain, J. W.: 1961, *Physics of Aurora and Airglow*, Academic Press, New York, Ch. 9.
Haymes, R. C.: 1971, *Introduction to Space Science*, John Wiley, New York, Ch. 5.
Silverman, S. M.: 1970, *Space Sci. Rev.* **11**, 341.

INDEX OF SUBJECTS

References in brackets denote illustrations

BIOGRAPHIES OF AUTHORS

Jacques M. *Beckers* was born in 1934 in Arnhem, The Netherlands, and studied at the University of Utrecht where he obtained his doctorate in 1964. He has done research at solar observatories in Australia, Germany and the USA and has been a staff member of the Sacramento Peak Observatory in New Mexico since 1962. His interest centres on the study of the dynamics of the solar chromosphere and of the properties of sunspots.

Anton *Bruzek* was born October 3, 1915 in Vienna, Austria, and studied at the University of Vienna where he obtained his PhD degree in 1939. In 1942 he joined the Fraunhofer Institut and did research work on the solar corona at mountain observatories at Wendelstein, Zugspitze and Kanzelhöhe. In 1953 he moved to Freiburg as Senior Scientist of the Fraunhofer Institut; he is engaged in observations and analysis of solar flares, filaments and chromospheric structures.

Helen *Dodson Prince* was born December 31, 1905 in Baltimore, Maryland. She obtained her PhD in 1933 from the University of Michigan. She was a member of the Faculty of Wellesley College 1933–1945 and Professor of Astronomy and Mathematics at Goucher College 1945–1950. Since 1947 until her retirement in 1976 she was Professor of Astronomy at the University of Michigan. She was first introduced to solar observations by M. and Mme. L. d'Azambuja in Meudon, France, in 1937 and 1938, and was appointed to McMath-Hulbert Observatory in 1946 where she participated for 30 years in the programs of observation and analysis of solar activity, solar flares, relationships between optical flares, radio frequency phenomena and energetic particles. Since 1961 she has been Associate Director of the observatory.

Christopher J. *Durrant* was born October 25, 1942 in London, UK. He made his undergraduate and graduate studies at Cambridge University where he obtained the PhD degree in 1969 with a thesis on magnetic A stars. He was Research Assistant at The Observatories, Cambridge 1969–1973 and is now staff member of the Fraunhofer Institut, Freiburg, Germany. His field of research is stellar atmospheric physics as typified in the Sun.

Adriaan D. *Fokker*, born in 1926, studied at the University of Leiden where he graduated in Astronomy in 1951. From 1952 to 1962 he worked in the field of solar radio astronomy and solar-terrestrial relationships in a research group of the Netherlands Postal and Telecommunication Services. In 1960 he took his doctor's degree with a thesis on metric solar radio noise storms. In 1962 he joined the Astronomical Institute of the University of Utrecht where he has held a post as associate professor since 1966.

John W. *Harvey* was born in Los Angeles, California on September 13, 1940. He received BA and MA degrees from the University of California and a PhD in 1969 from the University of Colorado. He has been employed by Lockheed Solar Observatory, High Altitude Observatory, Mt. Wilson Observatory and since 1969 has been on the staff of the Kitt Peak National Observatory. His interests centre on observational studies of solar magnetic and velocity fields and high resolution stellar spectroscopy and imaging.

Robert *Howard* received his doctorate from Princeton University in 1957. He was a Carnegie Fellow at the Mount Wilson and Palomar Observatories from 1957 to 1959 and was on the faculty of the University of Massachusetts (Amherst) from 1959 to 1961. Since 1961 he has been on the staff of the Mount Wilson and Palomar (later Hale) Observatories. His research interests include solar magnetic fields, solar activity, large-scale solar velocity fields, and the rotation of the Sun.

After obtaining her PhD in the Department of Astronomy, University College London, in 1965, Carole *Jordan* went as a Research Associate to the Joint Institute for Laboratory Astrophysics at the University of Colorado. She was then in the Science Research Council's Astrophysics Research Unit at Culham Laboratory for some nine years, but spent most of 1974 at the Centre for Astrophysics, Harvard University, as a Guest Investigator on Skylab data. Dr. Jordan is at present the Wolfson Tutorial Fellow in Natural Science at Somerville College, Oxford, and a Lecturer in the Department of Theoretical Physics at Oxford University. Her work has mainly concerned the interpretation of the solar EUV and X-ray spectrum, and has included studies of line identifications, methods of determining electron densities and models of the structure and energy balance of the quiet Sun and active regions.

Serge *Koutchmy* was born June 24, 1940 in Le Creusot, France. He made his graduate studies at Lomonossov University, Moscow, USSR and at the University of Orsay, France, and received his 'Doctorat d'Etat' in Physics and Astronomy from the University of Paris. Since 1967 he has been on the staff of the Institut d'Astrophysique, Paris, as 'researcher' of the French National Centre for Scientific Research (CNRS). Koutchmy studied the solar corona during five total solar eclipses and published more than 60 research papers concerning solar physics, planets, comets, interplanetary medium and instrumental astrophysics.

Marie-Josèphe *Martres* was born in 1924 in Paris and joined Meudon Observatory in 1955 as an assistant of Mme and M. d'Azambuja, and R. Michard. Since 1961 she has been in charge of the 'Cartes Synoptiques de la Chromosphère Solaire' and the 'Catalogues des Filaments et des Centres d'Activité'. Her research work deals with active region phenomena and evolution, and their relation to the solar corona. Some of her results have been used to establish criteria for short-term forecasts of solar activity. Mme Martres cooperated in a number of international solar projects and has been president of the Commission du Soleil of the Société Astronomique de France since 1973.

Vithalbhai L. *Patel* was born in India and became a naturalized US citizen in 1972. He obtained his BSc from the University of Baroda, MS from the University of Maryland and PhD from the University of New Hampshire in 1964. Before joining the University of Denver in 1966, where currently he is Professor of Physics, he did research at the University of New Hampshire, Rice University, and later at the University of California, Los Angeles and the University of Minnesota. He has published numerous research papers in the area of cosmic rays, geomagnetism, ionosphere, magnetic field and plasma in the magnetosphere and interplanetary space.

Ian W. *Roxburgh* was born in 1939 and studied at the University of Nottingham (BSc 1960) and the University of Cambridge (PhD, 1963). He has since worked at Cambridge, Kings College, London, the University of Sussex and Queen Mary College, London, with visits to the Max-Planck Institut, Munich, Goddard Space Flight Center, HAO and NCAR, Colorado, Caltech and the University of Virginia. He has been professor of Applied Mathematics at Queen Mary College, London since 1968. His interests embrace solar and stellar physics, gravitation and cosmology and meteorology.

Leif *Svalgaard* studied Geophysics at the University of Copenhagen, Denmark and joined the Danish Meteorological Institute 1965, after his graduation. Since 1972 he has been staff member of the Stanford Solar Observatory and of the Institute for Plasma Research at Stanford University. His present field of research comprises solar magnetic fields, their extensions into interplanetary space and their interactions with the geomagnetic field.

Einar A. *Tandberg-Hanssen* was born August 6, 1921 in Bergen, Norway. He graduated at Oslo University 1950 and did research work from 1951 to 1959 in Oslo, Pasadena, Cambridge (UK) and Boulder. He received his PhD degree from Oslo University in 1960. In 1961 he moved to USA and joined the High Altitude Observatory, Boulder, Col. as Senior Research Scientist; in 1974 he went to Marshall Space Flight Center, Alabama. He has been lecturing since 1959 in Oslo, Boulder and Alabama. His field of interest is characterised by the subjects of his two books: *Solar Activity* (Blaisdell Publishing Company, Waltham, Mass. 1967) and *Solar Prominences* (D. Reidel Publishing Company, Dordrecht 1974).

ASTROPHYSICS AND SPACE SCIENCE LIBRARY

1. C. de Jager (ed.), *The Solar Spectrum, Proceedings of the Symposium held at the University of Utrecht, 26–31 August, 1963.* 1965, XIV + 417 pp.
2. J. Ortner and H. Maseland (eds.), *Introduction to Solar Terrestrial Relations, Proceedings of the Summer School in Space Physics held in Alpbach, Austria, July 15–August 10, 1963 and Organized by the European Preparatory Commission for Space Research.* 1965, IX + 506 pp.
3. C. C. Chang and S. S. Huang (eds.), *Proceedings of the Plasma Space Science Symposium, held at the Catholic University of America, Washington, D.C., June 11–14, 1963.* 1965, IX + 377 pp.
4. Zdeněk Kopal, *An Introduction to the Study of the Moon.* 1966, XII + 464 pp.
5. B. M. McCormac (ed.), *Radiation Trapped in the Earth's Magnetic Field. Proceedings of the Advanced Study Institute, held at the Chr. Michelsen Institute, Bergen, Norway, August 16–September 3, 1965.* 1966, XII + 901 pp.
6. A. B. Underhill, *The Early Type Stars.* 1966, XII + 282 pp.
7. Jean Kovalevsky, *Introduction to Celestial Mechanics.* 1967, VIII + 427 pp.
8. Zdeněk Kopal and Constantine L. Goudas (eds.), *Measure of the Moon. Proceedings of the 2nd International Conference on Selenodesy and Lunar Topography, held in the University of Manchester, England, May 30–June 4, 1966.* 1967, XVIII + 479 pp.
9. J. G. Emming (ed.), *Electromagnetic Radiation in Space. Proceedings of the 3rd ESRO Summer School in Space Physics, held in Alpbach, Austria, from 19 July to 13 August, 1965.* 1968, VIII + 307 pp.
10. R. L. Carovillano, John F. McClay, and Henry R. Radoski (eds.), *Physics of the Magnetosphere, Based upon the Proceedings of the Conference held at Boston College, June 19–28, 1967.* 1968, X + 686 pp.
11. Syun-Ichi Akasofu, *Polar and Magnetospheric Substorms.* 1968, XVIII + 280 pp.
12. Peter M. Millman (ed.), *Meteorite Research. Proceedings of a Symposium on Meteorite Research, held in Vienna, Austria, 7–13 August, 1968.* 1969, XV + 941 pp.
13. Margherita Hack (ed.), *Mass Loss from Stars. Proceedings of the 2nd Trieste Colloquium on Astrophysics, 12–17 September, 1968.* 1969, XII + 345 pp.
14. N. D'Angelo (ed.), *Low-Frequency Waves and Irregularities in the Ionosphere. Proceedings of the 2nd ESRIN-ESLAB Symposium, held in Frascati, Italy, 23–27 September, 1968.* 1969, VII + 218 pp.
15. G. A. Partel (ed.), *Space Engineering. Proceedings of the 2nd International Conference on Space Engineering, held at the Fondazione Giorgio Cini, Isola di San Giorgio, Venice, Italy, May 7–10, 1969.* 1970, XI + 728 pp.
16. S. Fred Singer (ed.), *Manned Laboratories in Space. Second International Orbital Laboratory Symposium.* 1969, XIII + 133 pp.
17. B. M. McCormac (ed.), *Particles and Fields in the Magnetosphere. Symposium Organized by the Summer Advanced Study Institute, held at the University of California, Santa Barbara, Calif., August 4–15, 1969.* 1970, XI + 450 pp.
18. Jean-Claude Pecker, *Experimental Astronomy.* 1970, X + 105 pp.
19. V. Manno and D. E. Page (eds.), *Intercorrelated Satellite Observations related to Solar Events. Proceedings of the 3rd ESLAB/ESRIN Symposium held in Noordwijk, The Netherlands, September 16–19, 1969.* 1970, XVI + 627 pp.
20. L. Mansinha, D. E. Smylie, and A. E. Beck, *Earthquake Displacement Fields and the Rotation of the Earth, A NATO Advanced Study Institute Conference Organized by the Department of Geophysics, University of Western Ontario, London, Canada, June 22–28, 1969.* 1970, XI + 308 pp.
21. Jean-Claude Pecker, *Space Observatories.* 1970, XI + 120 pp.
22. L. N. Mavridis (ed.), *Structure and Evolution of the Galaxy. Proceedings of the NATO Advanced Study Institute, held in Athens, September 8–19, 1969.* 1971, VII + 312 pp.
23. A. Muller (ed.), *The Magellanic Clouds. A European Southern Observatory Presentation: Principal Prospects, Current Observational and Theoretical Approaches, and Prospects for Future Research, Based on the Symposium on the Magellanic Clouds, held in Santiago de Chile, March 1969, on the Occasion of the Dedication of the European Southern Observatory.* 1971, XII + 189 pp.

24. B. M. McCormac (ed.), *The Radiating Atmosphere. Proceedings of a Symposium Organized by the Summer Advanced Study Institute, held at Queen's University, Kingston, Ontario, August 3–14, 1970.* 1971, XI + 455 pp.
25. G. Fiocco (ed.), *Mesospheric Models and Related Experiments. Proceedings of the 4th ESRIN-ESLAB Symposium, held at Frascati, Italy, July 6–10, 1970.* 1971, VIII + 298 pp.
26. I. Atanasijević, *Selected Exercises in Galactic Astronomy.* 1971, XII + 144 pp.
27. C. J. Macris (ed.), *Physics of the Solar Corona. Proceedings of the NATO Advanced Study Institute on Physics of the Solar Corona, held at Cavouri-Vouliagmeni, Athens, Greece, 6–17 September 1970.* 1971, XII + 345 pp.
28. F. Delobeau, *The Environment of the Earth.* 1971, IX + 113 pp.
29. E. R. Dyer (general ed.), *Solar-Terrestrial Physics/1970. Proceedings of the International Symposium on Solar-Terrestrial Physics, held in Leningrad, U.S.S.R., 12–19 May 1970.* 1972, VIII + 938 pp.
30. V. Manno and J. Ring (eds.), *Infrared Detection Techniques for Space Research. Proceedings of the 5th ESLAB-ESRIN Symposium, held in Noordwijk, The Netherlands, June 8–11, 1971.* 1972, XII + 344 pp.
31. M. Lecar (ed.), *Gravitational N-Body Problem. Proceedings of IAU Colloquium No. 10, held in Cambridge, England, August 12–15, 1970.* 1972, XI + 441 pp.
32. B. M. McCormac (ed.), *Earth's Magnetospheric Processes. Proceedings of a Symposium Organized by the Summer Advanced Study Institute and Ninth ESRO Summer School, held in Cortina, Italy, August 30–September 10, 1971.* 1972, VIII + 417 pp.
33. Antonin Rükl, *Maps of Lunar Hemispheres.* 1972, V + 24 pp.
34. V. Kourganoff, *Introduction to the Physics of Stellar Interiors.* 1973, XI + 115 pp.
35. B. M. McCormac (ed.), *Physics and Chemistry of Upper Atmospheres. Proceedings of a Symposium Organized by the Summer Advanced Study Institute, held at the University of Orléans, France, July 31–August 11, 1972.* 1973, VIII + 389 pp.
36. J. D. Fernie (ed.), *Variable Stars in Globular Clusters and in Related Systems. Proceedings of the IAU Colloquium No. 21, held at the University of Toronto, Toronto, Canada, August 29–31, 1972.* 1973, IX + 234 pp.
37. R. J. L. Grard (ed.), *Photon and Particle Interaction with Surfaces in Space. Proceedings of the 6th ESLAB Symposium, held at Noordwijk, The Netherlands, 26–29 September, 1972.* 1973, XV + 577 pp.
38. Werner Israel (ed.), *Relativity, Astrophysics and Cosmology. Proceedings of the Summer School, held 14–26 August, 1972, at the BANFF Centre, BANFF, Alberta, Canada.* 1973, IX + 323 pp.
39. B. D. Tapley and V. Szebehely (eds.), *Recent Advances in Dynamical Astronomy. Proceedings of the NATO Advanced Study Institute in Dynamical Astronomy, held in Cortina d'Ampezzo, Italy, August 9–12, 1972.* 1973, XIII + 468 pp.
40. A. G. W. Cameron (ed.), *Cosmochemistry. Proceedings of the Symposium on Cosmochemistry, held at the Smithsonian Astrophysical Observatory, Cambridge, Mass., August 14–16, 1972.* 1973, X + 173 pp.
41. M. Golay, *Introduction to Astronomical Photometry.* 1974, IX + 364 pp.
42. D. E. Page (ed.), *Correlated Interplanetary and Magnetospheric Observations. Proceedings of the 7th ESLAB Symposium, held at Saulgau, W. Germany, 22–25 May, 1973.* 1974, XIV + 662 pp.
43. Riccardo Giacconi and Herbert Gursky (eds.), *X-Ray Astronomy.* 1974, X + 450 pp.
44. B. M. McCormac (ed.), *Magnetospheric Physics. Proceedings of the Advanced Summer Institute, held in Sheffield, U.K., August 1973.* 1974, VII + 399 pp.
45. C. B. Cosmovici (ed.), *Supernovae and Supernova Remnants. Proceedings of the International Conference on Supernovae, held in Lecce, Italy, May 7–11, 1973.* 1974, XVII + 387 pp.
46. A. P. Mitra, *Ionospheric Effects of Solar Flares.* 1974, XI + 294 pp.
47. S.-I. Akasofu, *Physics of Magnetospheric Substorms.* 1977, XVIII + 599 pp.
48. H. Gursky and R. Ruffini (eds.), *Neutron Stars, Black Holes and Binary X-Ray Sources.* 1975, XII + 441 pp.
49. Z. Švestka and P. Simon (eds.), *Catalog of Solar Particle Events 1955–1969. Prepared under the Auspices of Working Group 2 of the Inter-Union Commission on Solar-Terrestrial Physics.* 1975, IX + 428 pp.
50. Zdeněk Kopal and Robert W. Carder, *Mapping of the Moon.* 1974, VIII + 237 pp.
51. B. M. McCormac (ed.), *Atmospheres of Earth and the Planets. Proceedings of the Summer Advanced Study Institute, held at the University of Liège, Belgium, July 29–August 8, 1974.* 1975, VII + 454 pp.
52. V. Formisano (ed.), *The Magnetospheres of the Earth and Jupiter. Proceedings of the Neil Brice Memorial Symposium, held in Frascati, May 28–June 1, 1974.* 1975, XI + 485 pp.

53. R. Grant Athay, *The Solar Chromosphere and Corona: Quiet Sun.* 1976, XI + 504 pp.
54. C. de Jager and H. Nieuwenhuijzen (eds.), *Image Processing Techniques in Astronomy. Proceedings of a Conference, held in Utrecht on March 25–27, 1975,* XI + 418 pp.
55. N. C. Wickramasinghe and D. J. Morgan (eds.), *Solid State Astrophysics. Proceedings of a Symposium, held at the University College, Cardiff, Wales, 9–12 July 1974.* 1976, XII + 314 pp.
56. John Meaburn, *Detection and Spectrometry of Faint Light.* 1976, IX + 270 pp.
57. K. Knott and B. Battrick (eds.), *The Scientific Satellite Programme during the International Magnetospheric Study. Proceedings of the 10th ESLAB Symposium, held at Vienna, Austria, 10–13 June 1975.* 1976, XV + 464 pp.
58. B. M. McCormac (ed.), *Magnetospheric Particles and Fields. Proceedings of the Summer Advanced Study School, held in Graz, Austria, August 4–15, 1975.* 1976, VII + 331 pp.
59. B. S. P. Shen and M. Merker (eds.), *Spallation Nuclear Reactions and Their Applications.* 1976, VIII + 235 pp.
60. Walter S. Fitch (ed.), *Multiple Periodic Variable Stars. Proceedings of the International Astronomical Union Colloquium No. 29, Held at Budapest, Hungary, 1–5 September 1975.* 1976, XIV + 348 pp.
61. J. J. Burger, A. Pedersen, and B. Battrick (eds.), *Atmospheric Physics from Spacelab. Proceedings of the 11th ESLAB Symposium, Organized by the Space Science Department of the European Space Agency, held at Frascati, Italy, 11–14 May 1976.* 1976, XX + 409 pp.
62. J. Derral Mulholland (ed.), *Scientific Applications of Lunar Laser Ranging. Proceedings of a Symposium held in Austin, Tex., U.S.A., 8–10 June, 1976.* 1977, XVII + 302 pp.
63. Giovanni G. Fazio (ed.), *Infrared and Submillimeter Astronomy. Proceedings of a Symposium held in Philadelphia, Penn., U.S.A., 8-10 June, 1976.* 1977, X+226 pp.
64. C. Jaschek and G. A. Wilkins (eds.), *Compilation, Critical Evaluation and Distribution of Stellar Data. Proceedings of the International Astronomical Union Colloquium No. 35, held at Strasbourg, France, 19-21 August, 1976.* 1977, XIV+316 pp.
65. M. Friedjung (ed.), *Novae and Related Stars. Proceedings of an International Conference held by the Institut d'Astrophysique, Paris, France, 7-9 September, 1976.* 1977, XIV+228 pp.
66. David N. Schramm (ed.), *Supernovae. Proceedings of a Special IAU Session on Supernovae held in Grenoble, France, 1 September, 1976.* 1977, X+192 pp.
67. Jean Audouze (ed.), *CNO Isotopes in Astrophysics. Proceedings of a Special IAU Session held in Grenoble, France, 30 August, 1976.* 1977, XIII+195 pp.
68. Z. Kopal, *Dynamics of Close Binary Systems,* forthcoming.
69. A. Bruzek and C. J. Durrant (eds.), *Illustrated Glossary for Solar and Solar-Terrestrial Physics.* 1977, approx. 216 pp.
70. H. van Woerden (ed.), *Topics in Interstellar Matter.* 1977, VIII + 295 pp.

GPSR Compliance

The European Union's (EU) General Product Safety Regulation (GPSR) is a set of rules that requires consumer products to be safe and our obligations to ensure this.

If you have any concerns about our products, you can contact us on ProductSafety@springernature.com

In case Publisher is established outside the EU, the EU authorized representative is:

Springer Nature Customer Service Center GmbH
Europaplatz 3
69115 Heidelberg, Germany

Batch number: 10607320

Printed by Printforce, the Netherlands